MECHANICAL OPERATIONS

PARTICLES AND MOTION OF PARTICLES

MECHANICAL OPERATIONS

PARTICLES AND MOTION OF PARTICLES

D.C. SIKDAR

KHANNA BOOK PUBLISHING CO. (P) LTD.

Publisher of Engineering and Computer Books

4C/4344, Ansari Road, Darya Ganj, New Delhi-110002

Phone : 011-23244447-48 **Mobile:** +91-9910909320

E-mail : contact@khannabooks.com

Website : www.khannabooks.com

Dear Readers,

To prevent the piracy, this book is secured with HIGH SECURITY HOLOGRAM on the front title cover. In case you don't find the hologram on the front cover title, please write us to at contact@khannabooks.com or whatsapp us at +91-99109 09320 and avail special gift voucher for yourself.

Specimen of Hologram on front Cover title:

Moreover, there is a SPECIAL DISCOUNT COUPON for you with EVERY HOLOGRAM.

How to avail this SPECIAL DISCOUNT:

Step 1: Scratch the hologram
Step 2: Under the scratch area, your "coupon code" is available
Step 3: Logon to www.khannabooks.com
Step 4: Use your "coupon code" in the shopping cart and get your copy at a special discount
Step 5: Enjoy your reading!

Thanks,
Khanna Publishing House

Price: ₹ 345.00

ISBN: 978-93-5538-0500

Mechanical Operations: Particles and Motion of Particles
by **D.C. Sikdar**

First Edition: 2025

To view complete list of books, please scan the QR Code:

Published by:
Khanna Book Publishing Co. (P) Ltd.
CIN: U22110DL1998PTC095547

Visit us at: www.khannabooks.com
Write us at: contact@khannabooks.com

Printed in Bharat.

Contents

1

CHAPTER

Introduction

1.1 INTRODUCTION

The concept of Mechanical Operation was first developed by A.D. Little to explain the process of industrial chemistry in 1916. The idea of treating various process of chemical industry as a series of unit operations was put forward by W.H. Walker, W.K. Lewis and W.H. McAdam in their famous book "The principle of Chemical Engineering" in 1923. They have explained that the variety of chemical industry have processes which follow the same physical laws.

The performance of chemical engineering has to do with unit processes in which the raw materials are either changed or separated into useful products. The field of chemical engineering embraces physical or chemical changes of organic resources like agricultural, biological materials and inorganic resources like ores, coal and petroleum. The treatment of raw materials, chemical transformation and the separation of desired products are the few usual stages of any chemical process industry.

A chemical engineer plays a vital role in converting raw material into finished products through the process involving physical and chemical changes. A perfect chemical engineer is one who is skilled in development, design, construction and operation of industrial chemical plant in material undergoes a change.

1.2 MECHANICAL OPERATION

It is the unit operation that involves physically changing a material. Mechanical operation generally refers to change in size or shape. Mechanical operation also include separation of materials on the basis of physical properties like density, size and wettability. The operation may either be individual or may be a part of entire process. For example, grinding of stone for construction is an individual operation whereas the sugar industry in which sugar cane is first crushed by rotating mills before processing and centrifugation to produce sugar is a part of entire operation.

1.3 UNIT OPERATION

When the operations carried out in chemical process industry the physical changes in the system under consideration are regarded as unit operation. It involves a physical change or chemical transformation such as separation, crystallization, evaporation and filtration.

1.4 UNITS AND DIMENSIONS

The students of engineering will find that the data they uses are expressed in variety of different units. Therefore, the knowledge of all the system of units and its conversion factor is must to convert these data into a common system before proceeding with their calculations. Any physical quantities that can be measures is always expressed as product of numerical value and units thereof. For example, the distance between two city Bangalore and Hyderabad is expressed as 650 km. Here, 650 is the numerical value and km is a unit of distance. Physical quantities are divided into two types:

(*i*) Basic quantities

(*ii*) Derived quantities

1.4.1 Basic Quantities

Certain quantities are regarded as basic quantities such as length, mass, time and temperature. These are called dimensions or base units and are represented by the symbol L, M, T and θ. The basic quantities are represented by a system of units according to the system of measurement. Basically, the physical system representing the base unit differs in different system of units.

1.4.2 Derived Quantities

Certain quantities that obtained from the basic quantities are regarded as derived quantities such as area, velocity, pressure and energy. For example, consider velocity. It can be expressed as

$$V = \frac{d}{t} \quad \text{...(1.1)}$$

where, V = Velocity

d = Distance

t = time

It is composed of two base unit, length (L) and time (T). Hence, velocity is the derived quantity.

1.5 SYSTEM OF UNITS

The system of units composed of

(*i*) Base units

(*ii*) Derived units

(*iii*) Multiple units

1.5.1 Base Units

The units of basic quantities such as length, mass, time and temperature are called base units.

1.5.2 Derived Units

The units of the quantities obtained by multiplying and dividing base units are called derived units. For example, velocity = m/s.

1.5.3 Multiple Units

The units which are multiple or fraction of base units are called multiple units. For example, hour, minute and second.

The various system of units and the basic quantities associated with them are given in Table 1.1.

Table 1.1: System of units

Basic Quantity	System of Units				Dimensions
	CGS	MKS	FPS	SI	
Length	Centimetre (cm)	Metre (m)	Foot (ft)	Metre (m)	L
Mass	Gram (g)	Kilogram (kg)	Pound (lb)	Kilogram (kg)	M
Time	Second (s)	Second (s)	Second (s)	Second (s)	T
Temperature	Celsius (°C)	Celsius (°C)	Fahrenheit (°F)	Kelvin (K)	θ

There could be any number of derived quantities. Therefore, only commonly used derived quantities associated with their units are listed here in Table 1.2.

Table 1.2: Units of derived quantities

Derived Quantity	System of Units				Symbol	Dimensions
	CGS	MKS	FPS	SI		
Area	cm^2	m^2	ft^2	m^2	A	L^2
Volume	cm^3	m^3	ft^3	m^3 (l)	V	L^3
Density	g/cm^3	kg/m^3	lb/ft^3	kg/m^3	ρ	ML^{-3}
Velocity	cm/s	m/s	ft/s	m/s	v	LT^{-1}
Volumetric Flowrate	cm^3/s	m^3/s	ft^3/s	m^3/s (l/s)	Q	L^3T^{-1}
Mass Flowrate	g/s	kg/s	lb/s	kg/s	m	MT^{-1}
Force	$g.cm/s^2$ (dyn)	$kg.m/s^2$	$lb.ft/s^2$	$kg.m/s^2$	F	MLT^{-2}
Pressure	kgf/cm^2	kgf/m^2	lbf/ft^2	N/m^2 (Pa)	P	$ML^{-1}T^{-2}$
Work (Energy)	Erg	J	Btu	J (N.m)	W (E)	ML^2T^{-2}
Power	Erg/s	W	Btu/hr	W	P	ML^2T^{-3}
Viscosity	P (g/cm.s)	kg/m.s	lb/ft.hr	(Pa.s) kg/m.s ($N.s/m^2$)	μ	$ML^{-1}T^{-1}$
Thermal Conductivity	W/cm.°C	W/m.°C	Btu/hr.ft.°F	W/m.K	K	$MLT^{-3}\theta^{-1}$
Mass Velocity	$g/cm^2.s$	$kg/m^2.s$	$lb/ft^2.hr$	$kg/m^2.s$	G	$ML^{-2}T^{-1}$

1.6 CONVERSION OF UNITS

The conversion of units of particular quantity from one system of units to another system of units is done with the help of conversion factors when the quantity is expressed in terms of fundamental unit of mass, length, time and temperature. For example, to convert feet to metre, feet is need to multiply by the factor 0.3048. This factor is called conversion factor.

The mostly used conversion factors of some quantities are given in Table 1.3.

Table 1.3: Conversion factors

Length:	1 in = 2.54 cm
	1 ft = 0.3048 m
	1 mile = 1.6093 km
	1 yd = 0.9144 m

Area:	1 in^2 = 6.452 cm^2
	1 ft^2 = 0.929 m^2
	1 Acre = 4046.8 m^2
Volume:	1 in^3 = 16.386 cm^3
	1 ft^3 = 0.283 m^3
	1 lit. = 0.001 m^3
	1 lit. = 1000 ml
Capacity:	1 UK gal = 4.546 l
	1 US gal = 3.7854 l
	1 barrel = 0.1589 m^3
Mass:	1 lb = 0.4536 kg
	1 dyn = 0.00001 N
	1 grain = 0.0647 g
Density:	1 lb/ft^3 = 0.016 g/cm^3
Specific Volume:	1 ft^3/lb = 0.624 m^3/kg
Pressure:	1 lbf/in^2 = 0.0703 kgf/cm^2
	1 lbf/in^2 = 703 kgf/m^2
Energy:	1 Btu = 1055 J
	= 0.2522 kcal
	= 2.9307 × 10^{-4} kWhr
	1 lbf.ft = 1.3558 J
Power:	1 hp (electrical) = 746 W
	1 hp (metric) = 735.5 W
Heat Capacity:	1 Btu/lb.°F = 4186.7 J/kg°C
	1 Btu/lb°F = 1 kcal/kg°C
Temperature:	$°C = \frac{5}{9}(°F - 32)$
	$°F = \frac{9}{5}(°C) + 32$
	°R = °F + 459.67
	1 °R = 0.5556 K
Temperature interval difference:	1 °C = 1.8 °F = 1 K

1.7 FORCE

Force is defined as an external agent that is capable of changing the state of rest or motion of a particular body. It has magnitude and direction. It can be measured using a spring balance. The S.I. unit of force is Newton (N). It can be expressed by

$$F = m \cdot a \quad ...(1.2)$$

where, F = Force, N

m = Mass, kg

a = Acceleration, m/s^2

1 kgf = 9.806 N

1.8 PRESSURE

Pressure is defined as the force acting perpendicularly on the surface of the object per unit area. It can be expressed as

$$P = \frac{F}{A} \quad ...(1.3)$$

where, P = Pressure, N/m^2

F = Force, N

A = Area, m^2

The S.I. unit of pressure is N/m^2. The pressure is normally measured with the help of pressure gauge that registers the difference between the vessel and the local atmospheric pressure. This is called gauge pressure. The symbol for it is P_g. The gauge pressure does not indicate the true pressure. For obtaining the prevailing true pressure it is necessary to add atmospheric pressure in same units to the gauge pressure. This is called absolute pressure. The symbol for it is P_a. In general, if there is no suffix along with the symbol of pressure P, then it is considered as absolute pressure. It can be expressed as

$$P_a = P_g + P_{atm} \quad ...(1.4)$$

where, P_a = Absolute pressure, N/m^2

P_g = Gauge pressure, N/m^2

P_{atm} = Atmospheric pressure, N/m^2

The actual atmospheric pressure though varies from place to place its value at mean sea level is considered as 101325 N/m^2 or 1.033 kgf/cm^2. This is called normal atmospheric pressure. It can be expressed in terms of pressure head as

$$P_H = \frac{P_a}{\rho} \quad ...(1.5)$$

where, P_H = Pressure head, m

P_a = Absolute pressure, N/m^2

ρ = Density, kg/m^3

The subatmospheric pressure is refers to vacuum. It may be expressed as

$$P_a = P_{atm} - P_V \quad ...(1.6)$$

where, P_a = Absolute pressure, mmHg

P_{atm} = Atmospheric pressure, mmHg

P_V = Vacuum pressure, mmHg

Vacuum is usually expressed in unit mmHg or Torr. The most commonly used pressure heads are in terms of mercury and water columns as given below:

1 atm = 760 mmHg

= 10.33 mH_2O

= 101325 N/m^2 (Pa)

= 101.325 kPa

= 1.033 kfg/cm^2

= 1.013 bar

1.9 WORK

The product of the force acting on the body and distance travel by it in the direction of applied force is called work. It may be expressed as

$$W = F \times L \quad ...(1.7)$$

where, W = Work, J

F = Force, N

L = Distance, m

The S.I. unit of work or energy is Joule. The various form of energy present in the system are mechanical energy, chemical energy and thermal energy. Energy can be converted from one form to another form.

1.10 VOLUME

The amount of space that a substance occupies within a container is called volume. The S.I. unit of volume is m^3.

1.11 HEAT

Heat is the form of energy that is transferred between the system of high temperature to the system of low temperature. The S.I. unit of heat is Joule.

$$1 \text{ cal} = 4.184 \text{ J}$$

1.12 MATERIAL BALANCE

It is defined as the exact accounting of material entering and leaving the system. For unsteady state operations where the accumulation is changes with time the expression for material balance can be expressed as

$$\text{Input} = \text{Output} + \text{Accumulation} \quad ...(1.8)$$

For system at steady state conditions the time variation of system mass is equal to zero. Therefore, the equation (1.8) reduces to

$$\text{Input} = \text{Output} \quad ...(1.9)$$

1.13 ENERGY BALANCE

It is defined as the exact accounting of energy intake and energy expenditure of the system. For any unit operation:

$$\text{Energy input} = \text{Energy output} + \text{Energy accumulation} \quad ...(1.10)$$

For steady state operation:

$$\text{Energy input} = \text{Energy output} \quad ...(1.11)$$

1.14 MOLE FRACTION

It is the ratio of moles of individual substance to the total moles of the substances in the mixture. It can be expressed as:

$$\text{Mole fraction of } A = \frac{N_A}{N_A + N_B + N_C} \quad ...(1.12)$$

where N_A, N_B and N_C are the moles of substances *A*, *B* and *C* respectively. The sum of the mole fraction of substances in the mixture is 1. That means

$$N_A + N_B + N_C = 1 \quad ...(1.13)$$

Mole percent may be expressed as

$$\text{Mole percent of } A = \frac{N_A}{N_A + N_B + N_C} \times 100 \qquad \text{...(1.14)}$$

1.15 VOLUME FRACTION

It is the ratio of volume of individual substance to the total volume of the substance in the mixture. It can be expressed as

$$\text{Volume fraction of } A = \frac{V_A}{V_A + V_B + V_C} \qquad \text{...(1.15)}$$

where V_A, V_B and V_C are the volume of substances *A*, *B* and *C* respectively. The sum of the volume fraction of substances in the mixture is 1.

Volume percent may be expressed as

$$\text{Volume percent of } A = \frac{V_A}{V_A + V_B + V_C} \times 100 \qquad \text{...(1.16)}$$

1.16 WEIGHT FRACTION

It is the ratio of weight of individual substance to the total weight of the substances in the mixture. It can be expressed as:

$$\text{Weight fraction of } A = \frac{W_A}{W_A + W_B + W_C} \qquad \text{...(1.17)}$$

where W_A, W_B and W_C are the weight of substances *A*, *B* and *C* respectively. The sum of the weight fraction of substances in the mixture is 1. Weight percent may be expressed as:

$$\text{Weight percent of } A = \frac{W_A}{W_A + W_B + W_C} \times 100 \qquad \text{...(1.18)}$$

1.17 MOLECULAR UNIT

Molecular unit is often used instead of weight unit for simplification of calculation in problems of material balances involving chemical reactions.

For example, one kgmol of a substance is defined as the mass in kg of substance that is equal to numerically in its molecular weight. Thus, one kgmol of ethanol is equal to 46 kg of ethanol.

The number of moles of any substances can be obtained by the expression

$$N_A = \frac{W_A}{M_A} \qquad \text{...(1.19)}$$

where,

N_A = kgmol of *A*

W_A = Weight of *A*, kg

M_A = Molecular weight of *A*

1.18 IDEAL GAS LAW

Ideal gas law expresses the fact that the volume of a gas is directly proportional to the number of moles and temperature and inversely proportional to the pressure. Mathematically, the ideal gas law for *n* kgmol of gas can be expressed as

$$PV = nRT \qquad ...(1.20)$$

where,
P = Pressure, N/m^2
V = Volume, m^3
n = kgmol of gas
R = Universal gas constant, J/kgmol.K
T = Temperature, K

Let P_1, V_1 and T_1 are the pressure, volume and temperature of n kgmol of gas at state 1. Let P_2, V_2 and T_2 are the pressure, volume and temperature of n kgmol of gas at state 2.

Then,
$$P_1V_1 = nRT_1 \qquad ...(1.21)$$
$$P_2V_2 = nRT_2 \qquad ...(1.22)$$

Combining equation (1.21) and equation (1.22), we get

$$\frac{P_1V_1}{T_1} = \frac{P_2V_2}{T_2} \qquad ...(1.23)$$

The values of universal gas constant R changes with different system of units. The commonly used R values are given below:

R = 0.0820 l.atm/gmol.K
= 0.0820 m^3.atm/kgmol.K
= 8314 J/gmol.K
= 1.987 kcal/kgmol.K
= 1.987 Btu/lbmol.R
= 82.056 cm^3.atm/gmol.K
= 62362.7 cm^3.mmHg/gmol.K
= 10.7314 ft^3.

1.19 DALTON'S LAW

Dalton's law states that the total pressure exerted by a gaseous mixture is equal to the sum of the partial pressures of component gases. Thus, it expresses the additive nature of partial pressures. Mathematically, it can be expressed as:

$$P = p_A + p_B + p_C + ... \qquad ...(1.24)$$

where, P is the total pressure and p_A, p_B and p_C ... are the partial pressures of component gases A, B, C ... respectively.

1.20 PARTIAL PRESSURE

It is the pressure in the mixture of gases that would be exerted by that component gases if they were present in the same volume and temperature. It can be expressed as:

$$p_A = P \cdot x_A \qquad ...(1.25)$$

where,
p_A = Partial pressure of component A
P = Total pressure
x_A = Mole fraction of component A

1.21 VAPOUR PRESSURE

It may be defined as the absolute pressure at which the liquid and its vapour are in equilibrium at a given temperature. When the liquid is heated, evaporation takes place. During the evaporation two opposing process, the process of vapourization and the process of condensation occurs. As the vapourization proceeds, the pressure of vapourization increases and the rate of condensation also increases. The moment at which the rate of vapourization is equal to the rate of condensation, a sort of dynamic equilibrium exists and the vapour remains unchanged. The pressure exerted by the vapour at this condition is termed as vapour pressure. All liquids exhibit a definite vapour pressure at a given temperature.

1.22 AMAGAT'S LAW

It states that the total volume occupied by gaseous mixture is equal to the sum of the pure component volume of component gases. Mathematically, it can be expressed as:

$$V = V_A + V_B + V_C + \ldots \qquad \ldots(1.26)$$

where V is the total volume and V_A, V_B, V_C are the pure component volumes of the component gases A, B, C ... respectively.

1.23 PURE COMPONENT VOLUME

It is the volume that would be occupied by that component gases if they were present at the same temperature and pressure as of the mixture.

1.24 GASEOUS MIXTURES

The composition of component gases present in gas mixture is generally expressed in terms of volume percent. In case of gas mixture containing in a closed vessel, the molecules of each component gases are distributed throughout the entire volume of the container. The total pressure exerted by the entire mixture is equal to the sum of pressure exerted by each component gas.

The most useful form in which the laws of a mixture of ideal gases can be expressed is

$$\text{Volume\%} = \text{Mole\%} = \text{Pressure\%} \qquad \ldots(1.27)$$

1.25 RAOULT'S LAW

It states that the equilibrium partial pressure of component A is equal to the product of vapour pressure and mole fraction of component A in the liquid phase. It can be expressed as

$$p_A = P_A \cdot x_A \qquad \ldots(1.28)$$

where, p_A = Partial pressure of component A

P_A = Vapour pressure of pure liquid A

x_A = Mole fraction of component A in liquid phase

Since, y_A is the mole fraction of component A in the vapour phase is equal to the ratio of the partial pressure of A to the total pressure.

$$\therefore \qquad y_A = \frac{p_A}{P}$$

$$\Rightarrow \qquad y_A = \frac{P_A \cdot x_A}{P}$$

$$\Rightarrow \quad P \cdot y_A = P_A \cdot x_A = p_A \quad ...(1.29)$$

where, P = Total pressure

y_A = Mole fraction of A in the vapour phase

1.26 HENRY'S LAW

It state's that the partial pressure of solute gas is proportional to the mole fraction of that component in liquid phase. It can be expressed as:

$$p_A = H \cdot x_A \quad ...(1.30)$$

where, p_A = Partial pressure of component A

H = Henry's law constant

x_A = Mole fraction of A in liquid phase

Since, y_A is the mole fraction of component A in the vapour phase is equal to the ratio of the partial pressure of A to the total pressure.

$$\therefore \quad y_A = \frac{p_A}{P}$$

$$\Rightarrow \quad y_A = \frac{H \cdot x_A}{P}$$

$$p \cdot y_A = H \cdot x_A = p_A \quad ...(1.31)$$

where, P = Total pressure

y_A = Mole fraction of A in the vapour phase.

1.27 MECHANICAL LAW

It states that the applied force on a body is proportional to the product of mass and acceleration. It may be expressed as

$$F = k \cdot m \cdot a \quad ...(1.32)$$

where, F = Applied force on a body

m = Mass of body

a = Acceleration

k = Proportionality constant

1.28 DIMENSIONAL FORMULA

The equation obtained from physical quantity with its dimensional formulae are called dimensional formula. The dimensional formula helps in expressing physical quantities in terms of the fundamental quantities.

For example, the dimensional formula of force is

$$F = M \times L/T^2$$

$$F = MLT^{-2} \quad ...(1.33)$$

where, F = Force

M = Mass

L = Distance

T = Time

Each dimensional formula represent the dimension of mass, length, time and temperature. Following are the application of dimensional formula.

(*i*) To check the correctness of a physical relation.

(*ii*) To derive the relation between various physical quantities.

(*iii*) To find the dimension of constant in a given relation.

(*iv*) To convert value of physical quantity from one system of unit to another system.

1.29 DIMENSIONLESS EQUATION

All equations that have been derived directly from the basic fundamental laws consists of terms that have some dimensions because the basic fundamental laws themselves are used to define the derived quantities, no matter how complicated is the equation. An equation in which all terms have same dimensions is called dimensionally homogeneous equation.

When the dimensionally homogeneous equation is divided by only one of its term, the dimension of each term cancel and only magnitude remains. Such an equation is called dimensionless equation. For example, consider the Virial equation of state.

$$PV = RT + PV \qquad ...(1.34)$$

where, P = Absolute pressure, N/m^2

V = Specific volume, m^3/kg

R = Universal gas constant, J/kg.K

T = Absolute temperature, K

B = Virial coefficient, m^3/kg

Dividing equation (1.34) by RT, we get

$$\frac{PV}{RT} = 1+\frac{PB}{RT} \qquad ...(1.35)$$

It can be observed from the equation (1.35) that the dimensions of each term is cancelled and only the magnitude remains.

The significance of dimensionless equation is that it reduces the group containing physical variable. Hence, it reduces the time required to conduct the experiment as well as interpret the experimental data.

1.30 DIMENSIONLESS GROUP

A dimensionless group is the combination of dimensional or dimensionless quantities having zero overall dimension. Therefore, in a system of coherent units, it can be represented by a pure number.

In 1966, C.G. Stokes demonstrated the importance of dimensionless group now known as Reynolds number as a criterion for similarity of fluid behavior under different conditions. The use of dimensionless numbers in engineering and technology allows the important task of data reduction.

This means that a lot of experimental runs are avoided if data is correlated using appropriate dimensionless number.

A typical example of dimensionless group is the Reynolds number. It can be expressed as:

$$Re = \frac{DV\rho}{\mu} \qquad ...(1.36)$$

where, Re = Reynolds number

D = Diameter, L

V = Velocity, L/T
ρ = Density, M/L^3
μ = Viscosity, M/LT

$$\Rightarrow \quad Re = \frac{(L)(L/T)(M/L^3)}{M/LT}$$

$$\Rightarrow \quad Re = 1$$

Therefore, Re is a dimensionless number.

1.31 DIMENSIONAL EQUATION

The equations obtained by equalizing a physical quantity with its dimensional formulae are called dimensional equations. The dimensional equation helps in expressing physical quantities in terms of base or fundamental quantities.

For example, dimensional equation of acceleration may be expressed as:

$$[a] = \frac{\text{Velocity}}{\text{Time}}$$

$$\Rightarrow \quad [a] = \frac{L/T}{T}$$

$$[a] = M^{o}LT^{-2} \qquad \text{...(1.37)}$$

1.32 DIMENSIONALLY HOMOGENEOUS EQUATION

An equation is said to be dimensionally homogeneous when the dimensions of the term of an equation on the left hand side are equal to those on the right hand side. This is also called dimensional consistency of an equation.

For example, consider the enthalpy relation.

$$H = U + PV \qquad \text{...(1.38)}$$

where, H = Enthalpy, J/kg
U = Internal energy, J/kg
P = Absolute pressure, N/m^2
V = Specific volume, m^3/kg

The dimensional equation of enthalpy may be expressed as:

$$(MLT^{-2})(LM^{-1}) = (MLT^{-2})(LM^{-1}) + (MLT^{-2})(L^{-2})(L^3M^{-1}) \qquad \text{...(1.39)}$$

It can be observed from equation (1.39) that the dimensions of the term on left hand side are equal to those on the right hand side. Therefore, the enthalpy relation is dimensionally consistent.

1.33 DIMENSIONAL ANALYSIS

It is a method of determine the relationships between different physical quantities by using their basic fundamental properties length, mass, time and temperature. Dimensional analysis is a simple mathematical technique to determine the expression for dependent variables.

It is used in engineering and science research work for conducting model tests. The analysis can be carried out by the following methods:

(*i*) Rayleigh's method
(*ii*) Buckingham's-π method

1.33.1 Rayleigh's Method

This method of dimensional analysis is generally used when the expression contains maximum of three independent variables. If the number of independent variables is three or less, then it is very easy to determine the expression for the dependent variables. If the number of independent variable is four or more, then it is very difficult to determine the expression for the dependent variable.

The Rayleigh's method of dimensional analysis involves the following steps:

(*i*) Gather all the independent variables that are likely to influence the dependent variable.

(*ii*) If x is a variable that depends upon independent variables $x_1, x_2, x_3 \ldots x_n$, then the functional equation can be of the form

$$x = cf(x_1^a, x_2^b, x_3^c \ldots x_n^m) \quad \ldots(1.40)$$

(*iii*) Express each of the quantities in the equation in some base units in which the solution is required.

(*iv*) By using dimensional homogeneity obtain a set and determine the expression for the dependent variable.

1.33.2 Buckingham's-π Method

This method of dimensional analysis is generally used when the Rayleigh's method becomes difficult. This difficulties are overcome by using Buckingham's-π method of dimensional analysis while determining the expression for dependent variable.

In this method, there are 'n' number of independent and dependent variables and 'm' number of fundamental variables. The difference of these variables ($n - m$) will give rise to number of dimensionless terms. Each of this term is called Buckingham's-π term.

The Buckingham's-π method of dimensional analysis involves the following steps:

(*i*) List the number of independent and dependent variables 'n' involved in the problem such as fluid properties and geometry of the system.

(*ii*) Express each of the variables in terms of basic fundamental dimensions.

(*iii*) Determine the number of fundamental variables 'm'.

(*iv*) Determine the number of π-terms ($n - m$).

(*v*) Select the number of repeating variables, where the number required is equal to the number of reference dimensions.

(*vi*) Form a π-term by multiplying one of the non-repeating variable with the product of the repeating variable. Each raised to an exponent that will make the combination dimension.

(*vii*) Repeat the step for each of the remaining non-repeating variable.

(*viii*) Express the final form as relationship among the π-terms as

$$f(\pi_1, \pi_2, \pi_3 \ldots \pi_n) = 0 \quad \ldots(1.41)$$

1.34 USEFUL MATHEMATICAL METHOD

The mathematical technique of the calculations required for theory of mechanical operations which involves most elementary parts of calculus. There are two mathematical model, which are simple and useful in the treatment of this subject. They are:

(*i*) Graphical method of integration

(*ii*) Graphical treatment of exponential function.

1.34.1 Graphical Method of Integration

From the first principles of integral calculus we know that the value of a definite integral.

$$\int_{x_1}^{x_2} f(x) \cdot dx \qquad ...(1.42)$$

is the area bounded by the curve of $f(x)$ vs x, the ordinates $x = x_1$, $x = x_2$ and the x-axis. Any definite integral can therefore be evaluated numerically by plotting $f(x)$ against x and drawing the two vertical line through the values of x corresponding to the limits. Then, the area enclosed between the curve, the limits and x-axis can be measured as shown in Fig. 1.1.

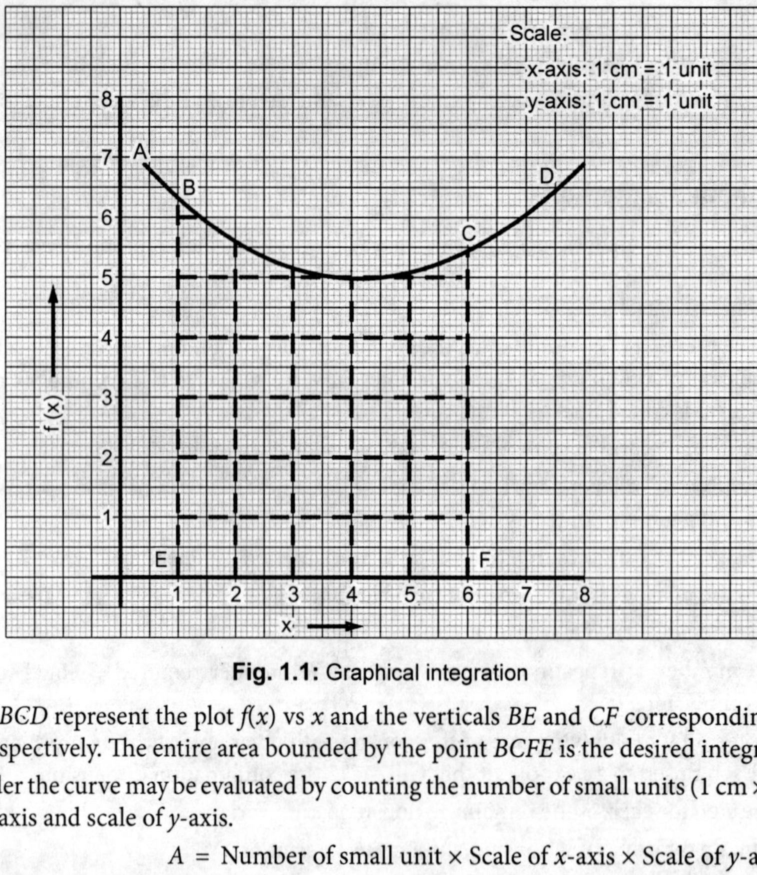

Fig. 1.1: Graphical integration

The curve *ABCD* represent the plot $f(x)$ vs x and the verticals *BE* and *CF* corresponding to the values of limit x_1 and x_2 respectively. The entire area bounded by the point *BCFE* is the desired integral.

The area under the curve may be evaluated by counting the number of small units (1 cm × 1 cm) multiplied by the scale of x-axis and scale of y-axis.

$$\therefore \qquad A = \text{Number of small unit} \times \text{Scale of } x\text{-axis} \times \text{Scale of } y\text{-axis}$$

1.34.2 Graphical Treatment of Exponential Function

It is found in many cases that experimental data involving variables x and y fit on expression of the type:

$$y = mx^n \qquad ...(1.43)$$

where, m and n are constant. Applying log on both side of the equation (1.43), we get

$$\log y = \log m + n \log x \qquad ...(1.44)$$

It is clear from the equation (1.44) that when $\log y$ is plotted against $\log x$, it will result in a straight line whose slope is n and intercept is $\log m$.

To plot a set of data by this method, we need to calculate the logarithm of the numerical values of the two variables, which is a tedious task. Therefore, instead of doing so, it is more convenient to plot these data on log-log graph. Log-log graph is coordinate graph on which the scales are logarithmic instead of being uniform.

The number of cycles in log-log graph is not fixed in either x-axis or y-axis. It varies depending on the size of the data. It is generally available in 2 cycles × 3 cycles and 3 cycles × 4 cycles. A typical log-log graph of 2 cycles × 2 cycles is shown in Fig. 1.2. It is observed from the graph that in the vertical axis as well as horizontal axis, the number moves from 1 to 9 for a cycle and again it starts with 1 for next cycle and so on. This is because of the fact that both x-axis and y-axis represent the logarithmic distance and there is no log of zero.

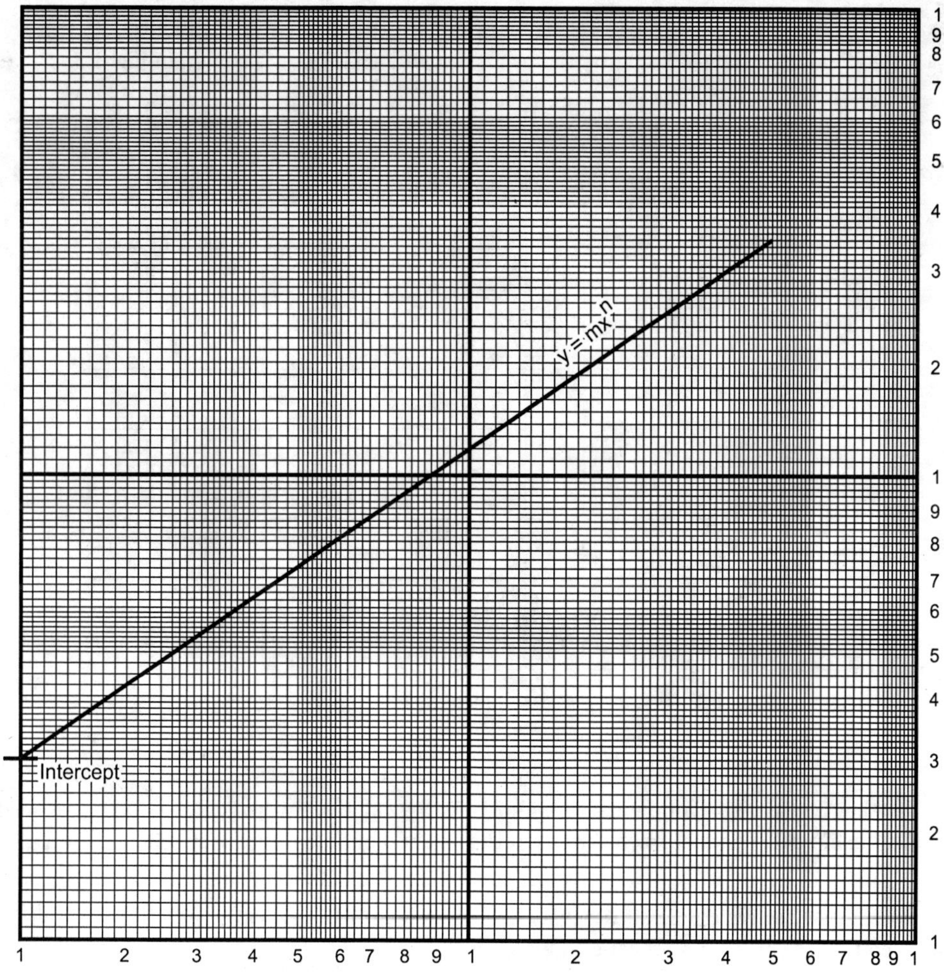

Fig. 1.2: Log-log graph

Log-log graph is useful when both of the variables being plotted to cover a large range of values. For example, in control engineering it is useful in plotting phase margin diagram for the determination of stability of the system.

1.37 LEAST SQUARE TECHNIQUE

It is a popular technique in which the overall solution of a system minimizes the sum of the square of the mistake made in the results of every single equation. Approximately, it gives the solution of a set of equations which consists of more equations than unknowns. This technique is available with two different version. Version one is called Ordinary Least Square (OLS) and the other version is called Weighted Least Square (WLS). Weighted least square is a most sophisticated technique which can regulate the importance of any observation in the final solution. Hence, its performance is better than OLS. The best fit in the sense of least square minimizes the difference between the observed values and the fitted values provided by the model equation.

For example, consider the best fit of a set of data points with a straight line function. A typical result of best fit of a set of data for a straight line function by least square technique is shown in Fig. 1.3.

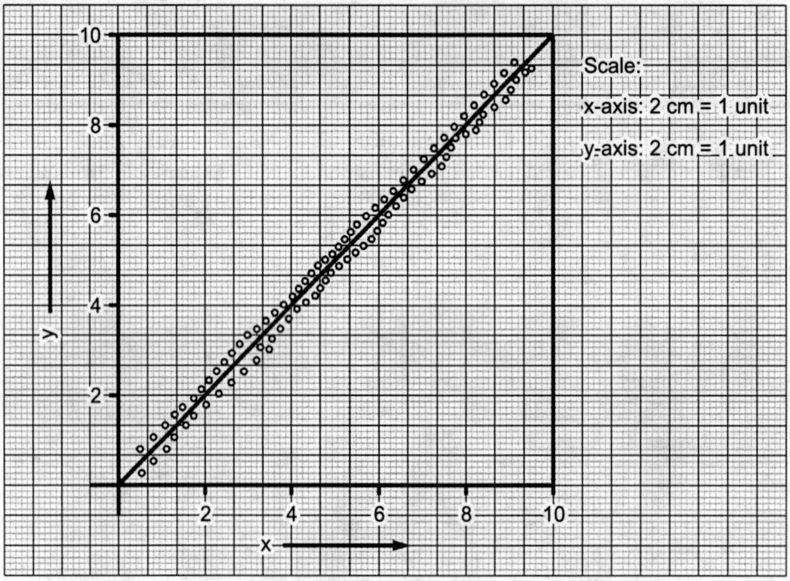

Fig. 1.3: Best fit for a straight line by least square technique

1.38 CURVE FITTING

Let us consider that we have experimental data comprising the value of two variables x and y. We need to find a possible relationship between these two variables. The method of obtaining the specific relation in the form

$$y = f(x) \qquad ...(1.45)$$

For the given set of experimental data to satisfy as accurately as possible is called curve fitting. This technique is generally used for the verification of experimental results.

Example 1.1. Express 1 atm in terms of pressure head of mercury column.

Solution: From the pressure head equation, we have

$$P_H = \frac{P_a}{\rho} \qquad ...(1.5)$$

where, P_H = Pressure head, m

P_a = 101325 N/m^2

$$\rho = 13600 \text{ kg/m}^3$$

$$\Rightarrow \quad P_H = \frac{101325}{13600 \times 9.808}$$

$$\Rightarrow \quad P_H = 0.760 \text{ mHg}$$

$$\Rightarrow \quad P_H = 760 \text{ mmHg}$$ **Ans.**

Example 1.2. Express 1 atm in terms of pressure head of water column.

Solution: From the pressure head equation, we have

$$P_H = \frac{P_a}{\rho} \quad ...(1.5)$$

where,

P_H = Pressure head, m

P_a = 101325 N/m^2

ρ = 1000 kg/m^3

$$\Rightarrow \quad P_H = \frac{101325}{1000 \times 9.808}$$

$$\Rightarrow \quad P_H = 10.22 \text{ mH}_2\text{O}$$ **Ans.**

Examples 1.3. Convert 1 kg/m.s to $N.s/m^2$.

Solution: From the conversion unit, we have

$$\Rightarrow \quad 1 \text{ N} = \text{kg.m/s}^2$$

$$\Rightarrow \quad 1 \text{ kg} = \text{N.s}^2\text{/m}$$

$$\Rightarrow \quad 1 \text{ kg/m.s} = 1 \times \left(\frac{\text{N.s}^2}{\text{m}}\right) \times \frac{1}{\text{ms}}$$

$$\Rightarrow \quad 1 \text{ kg/m.s} = 1 \text{ N.s/m}^2$$ **Ans.**

Example 1.4. Convert 1 paise to S.I. units.

Solution: From the conversion unit, we have

$$1 \text{ P} = 1 \text{ g/cm.s}$$

$$\Rightarrow \quad 1 \text{ P} = 1 \times \frac{1 \times 10^{-3}}{1 \times 10^{-2}}, \text{ kg/m.s}$$

$$\Rightarrow \quad 1 \text{ P} = 0.1 \text{ N.s/m}^2$$ **Ans.**

Example 1.5. Convert 1 CP to S.I. unit.

Solution: $$100 \text{ CP} = 1 \text{ P}$$

$$\Rightarrow \quad 1 \text{ CP} = \frac{1}{100} \text{ P}$$

$$\Rightarrow \quad 1 \text{ CP} = \frac{1}{100} \times 0.1 \text{ N.s/m}^2$$

$$\Rightarrow \quad 1 \text{ CP} = 0.001 \text{ N.s/m}^2$$ **Ans.**

Example 1.6. Convert 2500 l/s to m^3/s.

Solution: $$2500 \text{ l/s} = 2500 \times 10^{-3} \text{ m}^3\text{/s}$$

$$= 2.5 \text{ m}^3\text{/s}$$ **Ans.**

Example 1.7. Convert 2.5 atm into kPa.

Solution:

$$\begin{aligned} 2.5 \text{ atm} &= 2.5 \times 101.325 \text{ kPa} \\ &= 253.31 \text{ kPa} \end{aligned}$$

Ans.

Example 1.8. Convert 3.5 atm into bar.

Solution:

$$\begin{aligned} 3.5 \text{ atm} &= 3.5 \times 1.013 \text{ bar} \\ &= 3.546 \text{ bar} \end{aligned}$$

Ans.

Example 1.9. Convert 1 barrel into liter.

Solution:

$$\begin{aligned} 1 \text{ barrel} &= 0.1589 \text{ m}^3 \\ &= 0.1589 \times 1000 \text{ lit.} \\ &= 158.9 \text{ lit.} \end{aligned}$$

Ans.

Example 1.10. Show that 1 Btu/lb.°F = 1 kcal/kg.°C.

Solution:

$$\begin{aligned} 1 \text{ Btu/lb.}^\circ\text{F} &= 1 \times \frac{1055}{0.4536} \times \frac{1.8}{1} \text{ J/kg.}^\circ\text{C} \\ &= 4186.5 \text{ J/kg.}^\circ\text{C} \\ &= \frac{4186.5}{4.184} \text{ cal/kg.}^\circ\text{C} \\ &= 1 \text{ kcal/kg.}^\circ\text{C} \end{aligned}$$

Ans.

Example 1.11. Convert $1 \dfrac{\text{Btu}}{\text{hr.ft}^2.^\circ\text{R}}$ to $\dfrac{\text{w}}{\text{m}^2.\text{K}}$.

Solution:

$$\begin{aligned} 1 \frac{\text{Btu}}{\text{hr.ft}^2.^\circ\text{R}} &= 1 \times \frac{1055}{3600} \times \frac{1}{(0.3048)^2} \times \frac{1}{0.5556} \\ &= 5.678 \frac{\text{w}}{\text{m}^2.\text{K}} \end{aligned}$$

Ans.

Example 1.12. Convert $1 \dfrac{\text{Btu}}{\text{hr.ft.}^\circ\text{F}}$ to $\dfrac{\text{w}}{\text{m.K}}$.

Solution:

$$\begin{aligned} 1 \frac{\text{Btu}}{\text{hr.ft.}^\circ\text{F}} &= 1 \times \frac{1055}{3600} \times \frac{1}{0.3048} \times \frac{1.8}{1} \\ &= 1.7306 \frac{\text{w}}{\text{m.K}} \end{aligned}$$

Ans.

Example 1.13. Convert $1 \dfrac{\text{Btu}}{\text{hr.ft}^2.^\circ\text{F}}$ to $\dfrac{\text{w}}{\text{m}^2.\text{K}}$.

Solution:

$$\begin{aligned} 1 \frac{\text{Btu}}{\text{hr.ft}^2.^\circ\text{F}} &= 1 \times \frac{1055}{3600} \times \frac{1}{(0.3048)} \times \frac{1.8}{1} \\ &= 5.678 \frac{\text{w}}{\text{m}^2.\text{K}} \end{aligned}$$

Ans.

Example 1.14. Convert $1 \dfrac{\text{lbf.s}}{\text{ft}^2}$ to $\dfrac{\text{N.s}}{\text{m}^2}$.

Solution:

$$1\ \frac{\text{lbf.s}}{\text{ft}^2} = 1\times\frac{0.4536\times 9.806}{(0.3048)^2}$$

$$= 47.88\frac{\text{N.s}}{\text{m}^2} \qquad \textbf{Ans.}$$

Example 1.15. Convert $1\ \frac{\text{dyn.s}}{\text{cm}^2}$ to $\frac{\text{N.s}}{\text{m}^2}$.

Solution:

$$\frac{\text{dyn.s}}{\text{cm}^2} = 1\times\frac{0.00001}{1}\times\frac{10000}{1}$$

$$= 0.1\frac{\text{N.s}}{\text{m}^2} \qquad \textbf{Ans.}$$

Example 1.16. The power required by an agitator in a tank is the function of the following variables:

(*i*) Diameter of the impeller (D)

(*ii*) No. of rotation of the impeller per unit time (N)

(*iii*) Density of the liquid (ρ)

(*iv*) Viscosity of the liquid (μ)

Using Rayleigh's method of dimensional analysis, obtain the relation between the power and these variables.

Solution: From the Rayleigh's method of dimensional analysis, we have

$$P = f(D, N, \rho, \mu) \qquad \text{...(1.46)}$$

$$\Rightarrow \quad P = k[D^a, N^b, \rho^c, \mu^d] \qquad \text{...(1.47)}$$

The dimensions of each parameter in terms of fundamental base unit are:

Parameter	Units	Dimensions
Power (P)	N.m/s (kg.m^2/s^3)	ML^2T^{-3}
Diameter (D)	m	L
Rotation per unit time (N)	1/s	T^{-1}
Density (ρ)	kg/m^3	ML^{-3}
Viscosity (μ)	kg/m.s	$ML^{-1}T^{-1}$

Substituting dimensions in equation (1.47), we get

$$ML^2T^{-3} = k[(L)^a(T^{-1})^b(ML^{-3})^c(ML^{-1}T^{-1})^d \qquad \text{...(1.48)}$$

Comparing the power of M, L and T on both side of the equation (1.48), we get

$$M: 1 = c + d$$

$$L: 2 = a - 3c - d$$

$$T: -3 = -b - d$$

Solving in terms d:

$$c = 1 - d$$

$$b = 3 - d$$

$$2 = a - 3(1 - d) - d$$

$$2 = a - 3 + 3d - d$$

$$a = 5 - 2d$$

Substituting the value of a, b, c in equation (1.47), we get

$$P = k(D)^{5-2d}(N)^{3-d}(\rho)^{1-d}(\mu)^d$$

$$\Rightarrow \qquad P = k\left(\frac{D^5}{D^{2d}}\cdot\frac{N^3}{N^d}\cdot\frac{\rho^1}{\rho^d}\cdot\mu^d\right)$$

$$\Rightarrow \qquad \frac{P}{D^5N^3\rho} = k\left(\frac{\mu^d}{D^{2d}N^d\rho^d}\right)$$

$$\Rightarrow \qquad \frac{P}{D^5N^3\rho} = k\left(\frac{D^2N\rho}{\mu}\right)^{-d} \qquad \text{...(1.49)}$$

where, $\dfrac{P}{D^5N^3\rho}$ = Power number, N_P

$\dfrac{D^2N\rho}{\mu}$ = Impeller Reynolds number, N_{Re}

d = Constant, m

Therefore, the power number is a function of impeller Reynolds number to the power m. Infact, the power number N_P is also equal to Froude number, $N_F = \dfrac{DN^2}{g}$. **Ans.**

Example 1.17. The power required by an agitator in a tank is the function of the following variables:

(*i*) Diameter of the impeller (D)

(*ii*) No. of rotation of the impeller per unit time (N)

(*iii*) Density of the liquid (ρ)

(*iv*) Viscosity of the liquid (μ)

Using Buckingham's-π method of dimensional analysis, obtain the relation between the power and these variables.

Solution: From the Buckingham's-π method of dimensional analysis, we have

$$P = f(D, N, \rho, \mu) \qquad \text{...(1.50)}$$

$$\Rightarrow \qquad f(P, D, N, \rho, \mu) = 0 \qquad \text{...(1.51)}$$

Number of independent and dependent variables, $n = 5$.

Number of fundamental variables, $m = 3$.

Number of π-terms, $n - m = 5 - 3$

$= 2$

Therefore, $f(\pi_1, \pi_2) = 0$ ...(1.52)

The dimensions of each parameter in terms of fundamental base unit are:

Parameter	Units	Dimensions
Power (P)	N.m/s (kg.m^2/s^3)	ML^2T^{-3}
Diameter (D)	m	L
Rotation per unit time (N)	1/s	T^{-1}
Density (ρ)	kg/m^3	ML^{-3}
Viscosity (μ)	kg/m.s	$ML^{-1}T^{-1}$

Choosing D, N, ρ are repeating variable, we get

$$\pi_1 = D^{a_1} \cdot N^{b_1} \cdot \rho^{c_1} \cdot P \qquad ...(1.53)$$

$$\pi_2 = D^{a_2} \cdot N^{b_2} \cdot \rho^{c_2} \cdot \mu \qquad ...(1.54)$$

Substituting the dimensions of physical quantities in equation (1.53), we get

$$M^{\circ}L^{\circ}T^{\circ} = (L)^{a_1} \cdot (T^{-1})^{b_1} \cdot (ML^{-3})^{c_1} \cdot (ML^2T^{-3}) \qquad ...(1.55)$$

Comparing the power of M, L and T on both sides of the equation (1.55), we get

$$M : 0 = c_1 + 1$$

$$L : 0 = a_1 - 3c_1 + 2$$

$$T : 0 = -b_1 - 3$$

Solving above equations, we get

$$c_1 = -1$$

$$b_1 = -3$$

$$\Rightarrow \quad 0 = a_1 - 3(-1) + 2$$

$$\Rightarrow \quad 0 = a_1 + 5$$

$$\Rightarrow \quad a_1 = -5$$

Substituting the values of a_1, b_1 and c_1 in equation (1.53), we get

$$\pi_1 = (D)^{-5} \cdot (N)^{-3} \cdot (\rho)^{-1} \cdot P$$

$$\pi_1 = \frac{P}{D^5N^3\rho} \qquad ...(1.56)$$

Substituting the dimensions of physical quantities in equation (1.54), we get

$$M^{\circ}L^{\circ}T^{\circ} = (L)^{a_2} \cdot (T^{-1})^{b_2} \cdot (ML^{-3})^{c_2} \cdot ML^{-1}T^{-1} \qquad ...(1.57)$$

Comparing the power of M, L and T on both sides of the equation (1.57), we get

$$M : 0 = c_2 + 1$$

$$L : 0 = a_2 - 3c_2 - 1$$

$$T : 0 = -b_2 - 1$$

Solving above equations, we get

$$c_2 = -1$$

$$b_2 = -1$$

$$0 = a_2 - 3c_2 - 1$$

$$\Rightarrow \quad 0 = a_2 - 3(-1) - 1$$

$$\Rightarrow \quad 0 = a_2 + 3 - 1$$

$$\Rightarrow \quad a_2 = -2$$

Substituting the values of a_2, b_2 and c_2 in equation (1.54), we get

$$\pi_2 = (D)^{-2} \cdot (N)^{-1} \cdot (\rho)^{-1} \cdot \mu$$

$$\pi_2 = \frac{\mu}{D^2N\rho} \qquad ...(1.58)$$

Hence, $$f\left(\frac{P}{D^5N^3\rho}, \frac{\mu}{D^2N\rho}\right) = 0 \qquad ...(1.59)$$

$$\Rightarrow \qquad \frac{P}{D^5N^3\rho} = f\left(\frac{D^2N\rho}{\mu}\right) \qquad ...(1.60)$$

The group $\mu/D^2N\rho$ is arbitrarily inverted because the impeller Reynolds number N_{Re} is usually expressed in the form $D^2N\rho/\mu$.

Therefore, the power number N_P is a function of impeller Reynolds number N_{Re}. **Ans.**

Example 1.18. Calculate the volume occupied by 5 gmol of water vapour at 500°C and 10 atm by perfect gas law.

Solution: From the perfect gas law, we have

$$PV = nRT \qquad ...(1.20)$$

$$\Rightarrow \qquad V = \frac{nRT}{P}$$

where,

$$n = 5 \text{ gmol}$$

$$R = 82 \frac{\text{cm}^3.\text{atm}}{\text{gmol.K}}$$

$$T = 773 \text{ K}$$

$$P = 10 \text{ atm}$$

$$\Rightarrow \qquad V = \frac{5\times82\times773}{10}$$

$$\Rightarrow \qquad V = 31693 \text{ cm}^3$$

$$\Rightarrow \qquad V = 0.032 \text{ m}^3$$

Ans.

Example 1.19. During the experimental run in mechanical operation lab, it is observed that the difference of two limbs of mercury manometer is 1.2 cm. Calculate the absolute pressure of the system.

Solution: From the equation of absolute pressure, we have

$$P_a = P_g + P_{atm} \qquad ...(1.4)$$

where,

$$P_g = \rho gh$$

$$= 13600 \times 9.81 \times (1.2/100) \qquad ...(1.61)$$

$$= 1601 \text{ N/m}^2$$

$$P_{atm} = 101325 \text{ N/m}^2$$

$$\Rightarrow \qquad P_a = 1601 + 101325$$

$$\Rightarrow \qquad P_a = 102926 \text{ N/m}^2$$

Ans.

Example 1.20. Convert 102926 N/m^2 to kgf/cm^2.

Solution:

$$102926 \text{ N/m}^2 = 102926\times\frac{1}{9.81}\times\frac{1}{10^4}$$

$$= 1.049 \text{ kgf/cm}^2$$

Ans.

Example 1.21. During the experimental run in mass transfer lab, it is observed that the vacuum indicates by mercury manometer is 30 cm. Calculate the absolute pressure of the system.

Solution: From the equation of absolute vacuum, we have

$$P_a = P_{atm} - P_V$$

where, $P_{atm} = 760$ mmHg

$P_V = 30 \times 10$

$= 300$ mmHg

$$\Rightarrow \quad P_a = 760 - 300$$

$$\Rightarrow \quad P_a = 460 \text{ mmHg}$$ **Ans.**

Example 1.22. To prepare 5 kg, 3% solution of $CaCO_3$ in water, what is the amount of $CaCO_3$ and water must be added?

Solution: From the weight percent of A, we have

$$\frac{w_A}{w_A + w_B} \times 100 = 3\%$$

where, w_A = Weight of $CaCO_3$, g

$w_A + w_B$ = Weight of solution, 5000 g

$$\Rightarrow \quad \frac{w_A}{5000} \times 100 = 3$$

$$\Rightarrow \quad w_A = 150 \text{ g, } CaCO_3$$ **Ans.**

$$\text{Amount of water} = 5000 - 150$$

$$= 4850 \text{ g, water}$$ **Ans.**

Example 1.23. Show that Reynolds number N_{Re} is dimensionless.

Solution: From the equation of Reynolds number, we have

$$N_{Re} = \frac{DV\rho}{\mu} \quad ...(1.62)$$

Substituting the fundamental dimensions in equation (1.62), we get

$$\Rightarrow \quad N_{Re} = \frac{(L) \times (L/T) \times (M/L^3)}{(M/LT)}$$

$$\Rightarrow \quad N_{Re} = 1$$

Therefore, N_{Re} is dimensionless. **Ans.**

Example 1.24. Fit a straight line

$$y = ax + b \quad ...(1.63)$$

in the least square sense for the given data:

x	1	2	3	6	5	4
y	1	3	5	7	3	11

Solution: The normal equations for curve fitting.

$$y = ax + b \quad ...(1.63)$$

are given by

$$\Sigma y = a\Sigma x + nb \quad ...(1.64)$$

$$\Sigma xy = a\Sigma x^2 + b\Sigma x \quad ...(1.65)$$

Constructing the table based on the requirement, we have

x	y	xy	x^2
1	1	1	1
2	3	6	4
3	5	15	9
6	7	42	36
5	9	45	25
4	11	44	16
$\Sigma x = 21$	$\Sigma y = 36$	$\Sigma xy = 153$	$\Sigma x^2 = 91$

Substituting the above values in equation (1.64), we get

$$21a + 6b = 36 \qquad ...(1.66)$$

Substituting the above values in equation (1.65), we get

$$91a + 21b = 153 \qquad ...(1.67)$$

Multiplying equation (1.66) by 21 and equation (1.67) by 6, we get

$$441a + 126b = 756 \qquad ...(1.68)$$

$$546a + 126b = 918 \qquad ...(1.69)$$

$$\underline{- \quad\quad - \quad\quad -}$$

$$-105a = -162$$

$$a = 1.543$$

$$b = 0.6$$

Substituting the value of a and b in equation (1.63), we get

$$y = 1.543x + 0.6 \qquad ...(1.70)$$

This the best fit of a straight line.

Ans.

Example 1.25. Convert – 40°C into °F.

Solution: From the temperature relation, we have

$$\frac{°F - 32}{9} = \frac{°C}{5} \qquad ...(1.71)$$

$$\Rightarrow \quad °F - 32 = \frac{9}{5} \times (-40)$$

$$\Rightarrow \quad °F - 32 = -72$$

$$\Rightarrow \quad °F = -72 + 32$$

$$\Rightarrow \quad °F = -40$$ **Ans.**

Example 1.26. When 20 g NaOH is added in 81 ml of water. What is the mole percent of NaOH formed in solution?

Solution: From the equation of mole percent, we have

$$\text{Mole percent of } A = \frac{N_A}{N_A + N_B} \times 100 \qquad ...(1.72)$$

$$\text{Moles of NaOH,} \quad N_A = \frac{\text{Weight of NaOH in g}}{\text{Molecular weight of NaOH}}$$

$$= \frac{20}{40}$$

$$= 0.5$$

$$\text{Moles of } H_2O, \quad N_B = \frac{\text{Weight of } H_2O \text{ in g}}{\text{Molecular weight of } H_2O}$$

$$= \frac{81}{18}$$

$$= 4.5$$

$$\text{Mole percent of NaOH} = \frac{0.5}{0.5+4.5} \times 100$$

$$= 10\%$$

Ans.

REVIEW QUESTIONS

1.1 Briefly explain about mechanical operations.
1.2 What do you mean by unit operations? Explain.
1.3 Define unit.
1.4 Define dimensions.
1.5 What are basic quantities? Briefly explain.
1.6 What are derived quantities? Briefly explain.
1.7 What are multiple units. Give example.
1.8 What are the importance of conversion of units?
1.9 Define force.
1.10 Define pressure.
1.11 Explain the term absolute pressure.
1.12 Explain the term vacuum pressure.
1.13 Define work.
1.14 Define volume.
1.15 Define heat.
1.16 What do you mean by material balance. Explain with example.
1.17 Define energy balance.
1.18 Define mole fraction. Give example.
1.19 Define volume fraction. Give example.
1.20 Define weight fraction. Give example.
1.21 Explain the term molecular unit.
1.22 Define ideal gas law.
1.23 Define Dalton's law.
1.24 Define partial pressure.
1.25 Define vapour pressure.
1.26 Define Amagat's law.
1.27 Define pure component volume.
1.28 Briefly explain the term gaseous mixture.
1.29 Define Raoult's law.
1.30 Define Henry's law.
1.31 Explain the term mechanical law.
1.32 What do you mean by dimensional formula? Explain with example.
1.33 What do you mean by dimensionless equation? Explain with example.
1.34 What are the significance of dimensionless equation? Explain.
1.35 Write briefly on dimensionless group.
1.36 What do you mean by dimensional equation? Explain with example.
1.37 Write briefly on dimensionally homogeneous equation.
1.38 Explain dimensional analysis.

1.39 Explain the Rayleigh's method of dimensional analysis in detail.

1.40 Explain the Buckingham's-π method of dimensional analysis in detail.

1.41 Briefly explain the graphical method of integration.

1.42 Briefly explain the graphical treatment of exponential function.

1.43 What is least square technique? Explain.

1.44 Explain the significance of least square technique.

1.45 Briefly explain curve fitting and its applications in engineering.

1.46 Explain the usefulness of log-log graph.

1.47 What is dimensional consistency?

1.48 Define absolute viscosity.

1.49 Define dynamic viscosity.

EXERCISES

1.1 Convert 101325 N/m^2 to kgf/cm^2.

1.2 Convert 2 bar into atm.

1.3 Express 2 atm in terms of pressure head of mercury column.

1.4 Express 2.5 atm in terms of pressure head of water column.

1.5 Convert 2 $N.s/m^2$ into kg/m.s.

1.6 Convert 3 kg/m.s into poise.

1.7 Convert 50 poise into kg/m.s.

1.8 Convert 25 centi poise into kg/m.s.

1.9 Convert 5 atm into kPa.

1.10 Convert 10 barrel into liter.

1.11 Convert 5 kgf/m^2 to N/m^2.

1.12 Convert 2 $dyn.s/cm^2$ into kg/m.s.

1.13 Show that Prandtl number N_{pr} is dimensionless.

1.14 Convert – 40°F into °C.

1.15 The pressure drop ΔP in a pipe of diameter D and length l due to turbulent flow depends on the velocity V, viscosity μ, density ρ and roughness k. Using Buckingham's-π theorem obtain an expression for ΔP. **[Delhi University, Nov. 1982]**

1.16 Calculate the volume occupied by 3 gmol of water vapour at 300°C and 5 atm by perfect gas law.

1.17 The pressure drop ΔP in a pipe of diameter D and length l due to turbulent flow depends on the velocity V, viscosity μ, density ρ and roughness k. Using Rayleigh's theorem obtain an expression for ΔP.

1.18 During the experimental run in momentum transfer lab a mercury manometer is attached to a fluid flow system for the measurement of pressure drop. It is observed that the difference of two limb of mercury manometer is 1.5 cm. Calculate the absolute pressure of the system.

1.19 During the experimental run in mass transfer lab, it is observed that the vacuum indicates by mercury manometer is 35 cm. Calculate the absolute pressure of the system.

1.20 To prepare 1 litre, 3% solution of $CaCO_3$ in water, what is the amount of $CaCO_3$ and water must be added?

1.21 When 25 g NaOH is added in 225 ml of water. What is the weight fraction of NaOH in solution?

1.22 When 20 g NaOH is added in 63 ml of water. What is the mole percent of NaOH formed in solution?

1.23 Fit a straight line

$$y = ax + b$$

in the least square sense for the given data:

x	1	3	2	6	5	4
y	1	3	5	7	10	13

1.24 Assume air to contain 79% N_2 and 21% O_2 by volume. Convert these percentage into weight percent.

1.25 Express 1 atm in terms of pressure head of carbon tetrachloride.

❑❑❑

2 CHAPTER Particle Technology

2.1 INTRODUCTION

Particle technology generally deals with the Science and Technology related to the processing of particles. This is applied for the production, handling, modification and use of a wide variety of particulate materials. The materials may be either wet or dry in sizes ranging from nanometer to centimeter. Its scope ranges from industries which include chemical, petrochemical, agricultural, pharmaceutical, mineral and civil engineering to lab. The individual solid particles are characterized by their size, shape and density. Particles of homogeneous solids have the same density as the bulk material. Particles obtained by breaking up a composite solid such as metal bearing ore has various density usually different from the density of the bulk material. This will help the readers to grasp specific details of the applied technology.

2.2 PARTICLE SIZE

In physical science a particle is the small logical object which can be a scribed by several physical properties such as volume, density and mass. Anything that is composed of particles may be referred to as being particulate. In general, particle sizes are measured in micron (μ). A micron is equal to 1/1000 mm and a millimicron (mμ) is equal to 1/1000 micron. Usually the particle size is designated as the average diameter in microns, although some literature reports in terms of particle radius.

The particle diameter may be specified for any equidimensional particle. Particle that are not equidimensional *i.e.*, that are longer in one dimension than in others are sometime characterized by the second largest major dimension. By convension particle sizes are expected in different units depending on the size range involved.

The coarse particles are measured in millimeter, fine particles in terms of screen size and very fine in terms of micrometer or nanometer. Sometime, ultrafine particles are measured in terms of their surface area per unit mass usually in m^2/kg.

2.3 PARTICLE SHAPE

The shape of an individual particle is conveniently expressed in terms of particle sphericity ϕ_s, which is independent of particle size. For a spherical particle of diameter D_P, $\phi_s = 1$. For non-spherical particle the sphericity is defined by the relation.

$$\phi_s = \frac{6V_P}{D_P S_P} \qquad \text{...(2.1)}$$

where, ϕ_s = Sphericity

V_P = Volume of one particle

D_P = Equivalent diameter of the particle

S_P = Surface area of one particle

The equivalent diameter is sometimes are defined as the diameter of a sphere of equal volume. For fine granular materials however it is difficult to determine the exact volume and the surface area of the particle. Therefore, D_P is usually taken to be the nominal size of the particle on screen analysis or microscopic examination.

The surface area may be found from adsorption measurement or from the pressure drop in a bed of particles. For many crushed material ϕ_s is between 0.6 and 0.8. Particle rounded by absorption ϕ_s may be as high as 0.95.

2.4 SHAPE FACTOR

The shape of the particle of a solid is as important as particle size distribution. The shape does not vary with particle size, but no further concern is given to the particle shape. The definition of the shape factor is depend upon whether particle surface, volume or linear particle dimension is important to the application being concerned. A normal way the shape factor may be defined as the ratio of the particle property of a sphere having a diameter equal to the measured particle dimension. Thus, volume based shape factor may be expressed as

$$\text{Shape factor} = \frac{\text{Particle volume}}{\text{Volume of the sphere of same diameter}} \quad \text{...(2.2)}$$

The most commonly used shape factor is the sphericity, a surface volume shape factor. It is designated by the symbol ϕ_s.

Sphericity is defined as the ratio of the surface area of a sphere to the surface area of the particle. It may be expressed as

$$\phi_s = \frac{A_s}{S_P} \quad \text{...(2.3)}$$

where, ϕ_s = Sphericity

A_s = Surface area of a sphere

S_P = Surface area of the particle

$$\Rightarrow \quad \phi_s = \frac{\neq D_P^2}{S_P} \quad \text{...(2.4)}$$

$$\Rightarrow \quad \phi_s = \left(\frac{6V_P}{D_P}\right) \times \frac{1}{S_P}$$

$$\Rightarrow \quad \phi_s = \frac{6V_P}{D_P S_P} \quad \text{...(2.1)}$$

where, V_P = Volume of one particle

D_P = Equivalent diameter of one particle

An another method may be used to determine the shape factor. It is designated by the symbol η. It is the ratio of the surface of the particle to the surface of a shape of the same diameter. It may be expressed as

$$\eta = \frac{S_P}{6V_P/D_P} \quad \text{...(2.5)}$$

where, η = Shape factor

S_P = Surface area of the particle

V_P = Volume of one particle

D_P = Equivalent diameter of one particle

The diameter is usually taken as the mean screen opening and specific surface area is the surface area per unit mass of material.

Shape factor are dimensionless quantities used in image analysis and microscopy that numerically describe the shape of a particle independent of its size.

2.5 PARTICLE SIZE ANALYSIS

Particle size analysis is a laboratory technique which determines the size range of the particles in a given sample. The particle size analysis can be applied to solid materials, suspensions, emulsion and even aerosols. Some particle sizing method can be used for a wide range of sample, but some can only be used for specific applications. It is quite important to select the most suitable method for different sample as different method can produce quite different results for the same materials. Particle size analysis is a very important laboratory test and is used for quality control in many different industries such as milling and grinding. Particle size is a critical factor in determining the efficiency of manufacturing processes and performance of the product. There are number of methods of particle size analysis but most particle size determination are made by screen analysis when the particles are within the size range. Usually, for carrying out particle size analysis of a sample the standard screens of either Tyler standard screen series or US Sieve series are used.

The screen coarser than 200 mesh have their mesh and wire diameter so adjusted that the area of opening in any one screen is approximately twice the area of the opening in next consecutive screen. This means that the ratio of linear size of opening in any screen to that in the next linear screen is $\sqrt{2}$. Thus, it can be observed from the Tyler series of standard screen that a 150 mesh screen will have an opening of 0.105 mm with wire diameter of 0.063 mm.

The US Sieve series which uses the Tyler series 200 mesh standard but differing slightly in other size is also widely used. Normally, 200 mesh screen is used as finest one, even though several finer screen down to 400 mesh are available. These finer screens are seldom used except in laboratory investigation. That is because of the reason that the finer screens are expensive, delicate and have serious problem of binding. The seive either Tyler standard or US seive series are used to determine the effectiveness of screening equipment and work performance of crushing and grinding machinery.

2.6 MIXED PARTICLE SIZE ANALYSIS

In a sample of uniform particles of diameter D_P the volume of the particle is expressed as

$$V_P = \frac{m}{\rho_P} \quad \text{...(2.6)}$$

where, V_P = Volume of the particles

m = Mass of the sample

ρ_P = Density of the particle

Since the volume of one particle is V_P, the number of particles in the sample may be expressed as

$$N = \frac{m}{V_P \rho_P} \quad \text{...(2.7)}$$

The total surface area of the particle in the sample can be expressed as

$$A = N \cdot S_P \quad \text{...(2.8)}$$

where, A = Total surface area of the sample

N = Number of particles

S_P = Specific surface area of the particle

From the equation of sphericity, we have

$$\phi_s = \frac{6V_P}{D_P \cdot S_P} \qquad \text{...(2.1)}$$

$$\Rightarrow \qquad S_P = \frac{6V_P}{\phi_s \cdot D_P} \qquad \text{...(2.9)}$$

Substituting equation (2.7) and equation (2.9) in equation (2.8), we get

$$A = \frac{m}{V_P \cdot \rho_P} \times \frac{6V_P}{\phi_s \cdot D_P}$$

$$\Rightarrow \qquad A = \frac{6m}{\phi_s \cdot D_P \cdot \rho_P}, m^2 \qquad \text{...(2.10)}$$

Using this equation total surface area of the particle in the sample may be evaluated.

2.7 SCREEN

It is a device used to separate the solids into two or more product on the basis of their size and this process is called screening. The minus particle pass through the screen at a diminishing rate until all particles closest to the opening size have been separated out. The action of screening is aided when screen is subjected to some kind of motion either reciprocating or gyratory in the horizontal plane. The time duration of the shaking of screen to reach this stage is roughly proportional to the amount of the material on the screen.

The performance of the screen mainly depends on three parameters namely:

(*i*) Motion

(*ii*) Inclination

(*iii*) Screening media

The most commonly used measure of screen efficiency is the cumulative weight of material that has passed the screen in any time interval compared to the total weight of undersize in the feed expressed in a percentage. This can be reversed when the oversize is the product to be recovered than efficiency is the weight percent of material in the screened oversize fraction compared to the total weight of oversize in the feed.

The process of screening falls under two categories namely:

(*i*) Screening by stratification

(*ii*) Screening by free fall

In the case startification screening is carried out by building up a material bed on a screen deck. The material get stratified when the motion of the screen reduces the internal friction in the sample material. This means that the finer particles can pass between the larger ones giving a sharp separation.

In case of screening by free fall the double inclination is used for stratification. During the free fall no particle layer can build up on screen deck and the particles are sized directly through the screening media giving a higher capacity with less sharpness in separation.

Sizing of screen is a time consuming process since various factors influences it and hence it is usually done by the specialists. Selection of the correct size and type of screen is very important, equally important is the selection of the screening media. This refers not only to a correct aperture related to the cut size, but also to the wear in operation of these screens.

The general rule for the minimum thickness of the screen panel is that it should be one fourth of the maximum feed size. In case the screen panel is thicker, the capacity and the accuracy reduces and service life and the blinding tendency increases. Therefore, the panel thickness is not to exceed the required product size.

A typical screen generally used in mechanical operation lab is shown in Fig. 2.1.

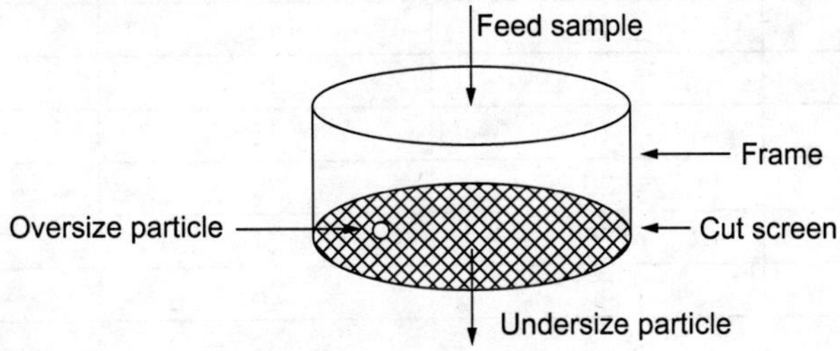

Fig. 2.1: Single screen

2.8 IDEAL SCREEN AND ACTUAL SCREEN

The objective of a screen is to accept a feed containing a mixture of particles of various sizes and separate it into two fractions, an underflow that is passed through the screen and an overflow that is rejected by the screen. Either one or both of these streams may be the product.

An ideal screen would sharply separate the feed sample in such a way that the smallest particle in the overflow would be just larger than the largest particle in the underflow. Such an ideal separation defines a cut diameter D_{pc} that marks point of separation between the fractions. Usually, D_{pc} is chosen to be equal to the mesh opening of the screen.

An actual screen do not give a perfect separation about the cut diameter. The closest separation are obtained with spherical particle on standard testing screen, but even here there is an overlap between the smallest particle in the overflow and the largest one in the underflow. The overlap is specially pronounced when the particles are needle like or fibrous or where the particles tend to aggregate into clusters that act as larger particles. Commercial screen usually give poorer separation than testing screens of the same mesh opening operating on the same samples.

2.9 US SIEVE SERIES AND TYLER STANDARD MESH EQUIVALENT

Two scales that are used to classify particle sizes are the US Sieve series and Tyler standard mesh equivalent. Sometimes Tyler standard mesh equivalent also called Tyler standard sieve series. The US sieve sizes differ from Tyler screen size, in that they are arbitrary number. The most common mesh opening for these scales are given in the Table 2.1. It provides an indication of particle size.

Table 2.1: US sieve series and Tyler standard mesh equivalent

Sl. No.	US sieve size no.	Tyler standard mesh equivalent no.	Opening (mm)
1	4	4	4.76
2	5	5	4.00
3	6	6	3.36
4	7	7	2.83
5	8	8	2.38

Sl. No.	US sieve size no.	Tyler standard mesh equivalent no.	Opening (mm)
6	10	9	2.00
7	12	10	1.68
8	14	12	1.41
9	16	14	1.19
10	18	16	1.00
11	20	20	0.841
12	25	24	0.707
13	30	28	0.595
14	35	32	0.500
15	40	35	0.420
16	45	42	0.354
17	50	48	0.297
18	60	60	0.250
19	70	65	0.210
20	80	80	0.177
21	100	100	0.145
22	120	115	0.125
23	140	150	0.105
24	170	170	0.088
25	200	200	0.074
26	230	250	0.063
27	270	270	0.053
28	325	325	0.044
29	400	400	0.037

2.10 SCREEN ANALYSIS

The fine particles are generally specified according to their screen analysis. A screen analysis of a material is carried out by using testing devices. Screen analysis is a practice used in chemical engineering to assess the particle size distribution of a granular material by allowing it to pass through a set of standard screen arranged serially in a stack in such a way that the coarsest of the screen is at the top and the finest of the screen is at the bottom. An analysis is carried out by placing the sample on the top of the screen and shaking the stack either manually or mechanically for a definite length of time. The material retained on each screen is removed and weighed. For reporting the screen analysis the amount of material retained on each screen is expressed as the weight fraction of the total mass of the sample.

The screen analysis of a sample may be reported either in tabular form or as graph. In general, the results of a screen analysis reported in tabular form to show the weight fraction of material retained on each screen as a function of the mesh size. As the particles retained on any one screen are passed through the screen immediately above it. Therefore, two numbers are needed to specify the size, one for the screen through which the fraction passes and the other for the screen on which the fraction is retained. Hence, the notation 35/48 means the material passes through the screen of mesh number 35 and retained on the screen of mesh number 48. The analysis reported in this manner in a tabular form is called differential analysis.

The materials that is retained on the screen is the oversize (+) and the material that passes through it is the undersize (–) of that screen. Thus a – 35 + 48 fraction means, the fraction of the material that passes through a 35 mesh but is retained on a 48 mesh screen. The average particle size of the material retained on any particular screen is the arithmetic mean of two screen opening used to obtain the fraction. A typical differential screen analysis is shown in Table 2.2.

Table 2.2: Differential screen analysis

Sl. No.	Mesh No.	Screen opening (microns)	Average particle size (microns)	Weight fraction retained
1	6	3360	2870	0.015
2	8	2380	2030	0.237
3	10	1680	1435	0.296
4	14	1190	1015	0.219
5	20	841	718	0.108
6	28	595	507	0.058
7	35	420	358	0.024
8	48	297	253	0.016
9	65	210	179	0.010
10	100	149	127	0.008
11	150	105	89	0.003
12	200	74	68	0.002
13	250	63	53	0.004
14	Pan	—	—	—

The differential screen analysis data given in Table 2.2 can be reported in graph form as shown in Fig. 2.2

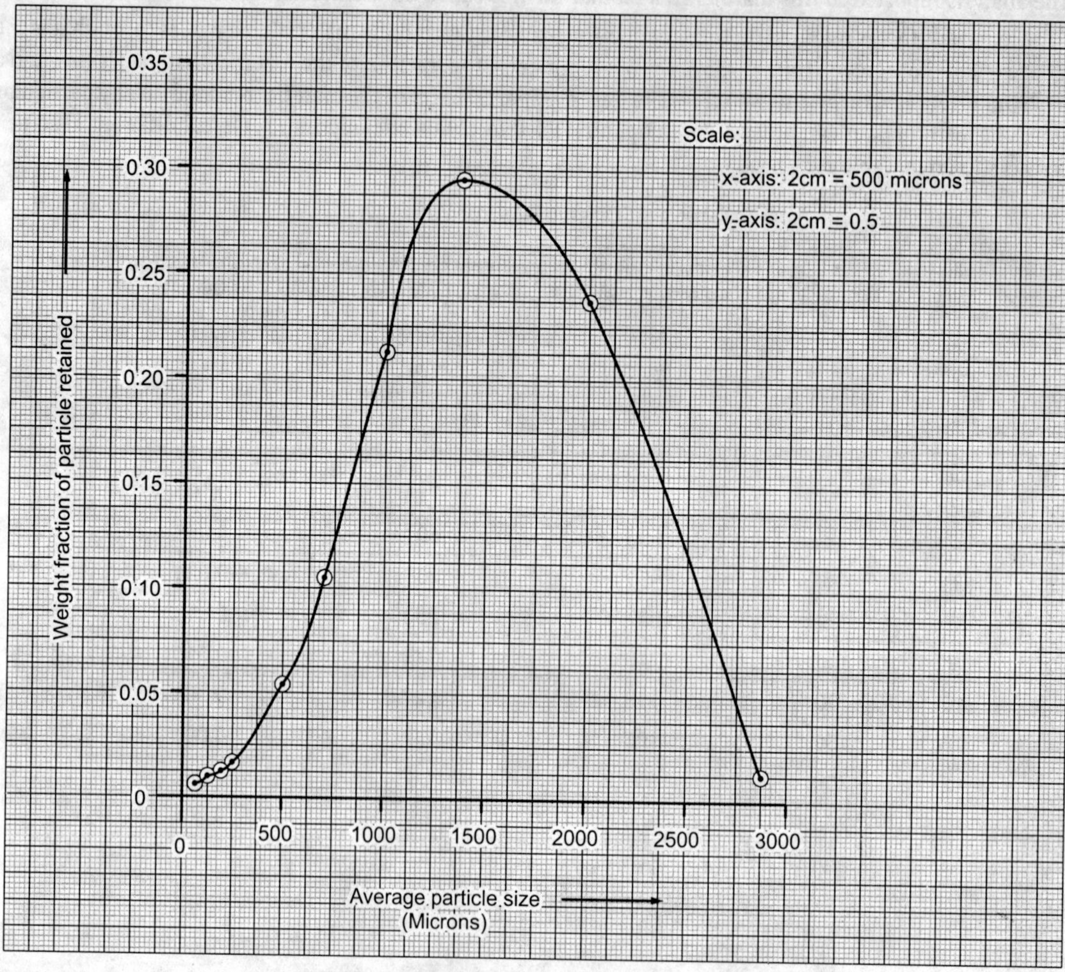

Fig. 2.2: Differential screen analysis of sample

One another method of reporting the information of screen analysis is cumulative analysis. A cumulative analysis is obtained from the differential analysis data by adding cumulatively the individual mass fractions of material that retained on each screen starting from the largest mesh to smallest mesh and plotting the cumulative mass fraction retained (ϕ) against screen opening. A cumulative analysis may also be obtained from the differential analysis data by incorporating cumulative mass fraction that passes ($1 - \phi$) through the screen. Cumulative analysis is more precise than those based on differential analysis. Assumption in cumulative analysis is that all particles in a single fraction are equal in size is not needed. A typical cumulative screen analysis obtained from the differential screen analysis data is shown in Table 2.3

Table 2.3: Cumulative screen analysis

Sl. No.	Mesh No.	Screen opening microns	Weight fraction retained	Cumulative mass fraction retained (ϕ)	Cumulative mass fraction passing ($1-\phi$)
1	6	3360	0.000	0.000	1.000
2	8	2380	0.015	0.015	0.985
3	10	1680	0.237	0.252	0.748
4	14	1190	0.296	0.548	0.452
5	20	841	0.219	0.767	0.233
6	28	595	0.108	0.875	0.125
7	35	420	0.058	0.933	0.067
8	48	297	0.024	0.957	0.043
9	65	210	0.016	0.973	0.027
10	100	149	0.010	0.983	0.017
11	150	105	0.008	0.991	0.009
12	200	74	0.003	0.994	0.006
13	250	63	0.002	0.996	0.004
14	Pan	—	0.004	1.000	0.000

The cumulative screen analysis data given in Table 2.3 can be reported in graph form as shown in Fig. 2.3.

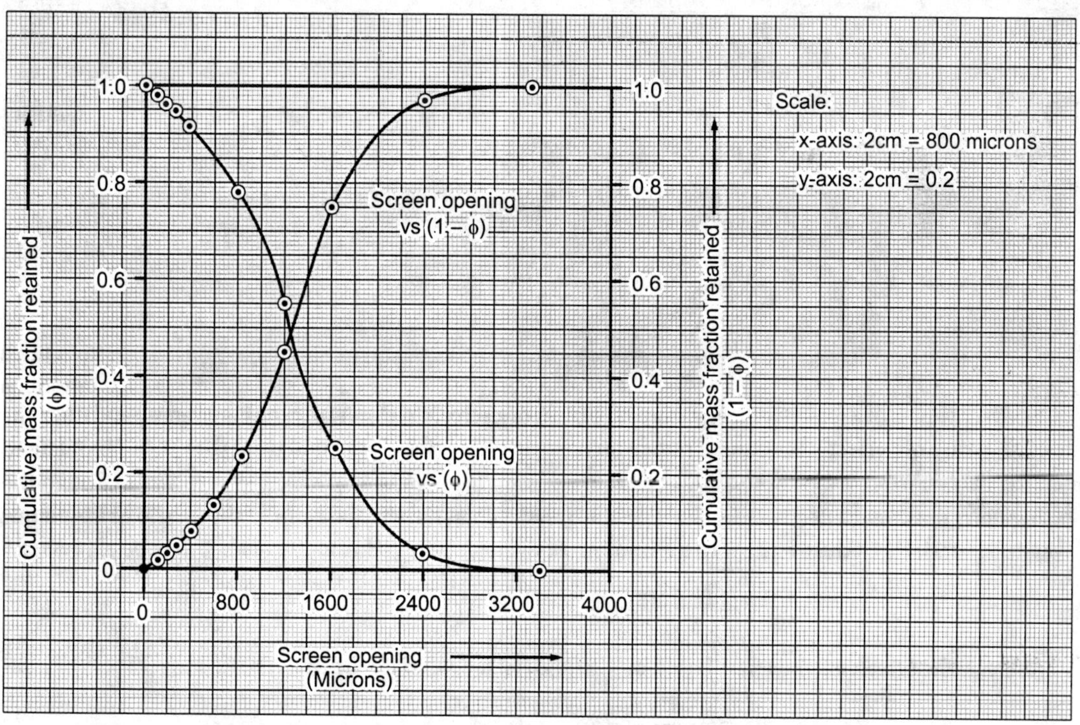

Fig. 2.3: Cumulative screen analysis of sample

2.11 MATERIAL BALANCE OVER SCREEN

Consider a sample consists material A and B is feed for screening, where A is oversize and B is the undersize material.

Let F, D and B be the mass flow rates of feed, overflow and underflow respectively.

Let x_F, x_D and x_B be the mass fractions of material A in the feed, overflow and underflow respectively. Therefore, the mass fraction of material B in the feed, overflow and underflow are $(1 - x_F)$, $(1 - x_D)$ and $(1 - x_B)$ respectively.

The overall material balance over a screen is expressed as

$$F = D + B \qquad \text{...(2.11)}$$

From the material balance of A over the screen, we get

$$F \cdot x_F = D \cdot x_D + B \cdot x_B \qquad \text{...(2.12)}$$

From the equation (2.11), we have

$$D = F - B \qquad \text{...(2.13)}$$

Substituting equation (2.13) in equation (2.12), we get

$$F \cdot x_F = (F - B)x_D + B \cdot x_B$$

$$\Rightarrow \quad F \cdot x_F = F \cdot x_D - B \cdot x_D + B \cdot x_B$$

$$\Rightarrow \quad F \cdot x_F - F \cdot x_D = B \cdot x_B - B \cdot x_D$$

$$\Rightarrow \quad F \cdot x_D - F \cdot x_F = B \cdot x_D - B \cdot x_B$$

$$\Rightarrow \quad F(x_D - x_F) = B(x_D - x_B)$$

$$\Rightarrow \quad \frac{B}{F} = \frac{(x_D - x_F)}{(x_D - x_B)} \qquad \text{...(2.14)}$$

From the equation (2.11), we have

$$B = F - D \qquad \text{...(2.15)}$$

Substituting equation (2.15) in equation (2.12), we get

$$F \cdot x_F = D \cdot x_D + (F - D) \cdot x_B$$

$$\Rightarrow \quad F \cdot x_F = D \cdot x_D + F \cdot x_B - D \cdot x_B$$

$$\Rightarrow \quad F \cdot x_F - F \cdot x_B = D \cdot x_D - D \cdot x_B$$

$$\Rightarrow \quad F(x_F - x_B) = D(x_D - x_B)$$

$$\Rightarrow \quad \frac{D}{F} = \frac{(x_F - x_B)}{(x_D - x_B)} \qquad \text{...(2.16)}$$

The effectiveness of a screen is the measure of success of a screen in closely separating oversize and undersize materials. In case of perfectly functional screen all the oversize material A would be in the overflow and all the undersize material B would be in the under flow.

Therefore, the screen effectiveness based on the oversize material may be defined as the ratio of the amount of oversize material A in the overflow to the amount of oversize material A in the feed. It may be expressed as

$$E_A = \frac{D \cdot x_D}{F \cdot x_F} \qquad \text{...(2.17)}$$

Similarly, the screen effectiveness based on the undersize material may be expressed as

$$E_B = \frac{B(1 - x_B)}{F(1 - x_F)} \qquad \text{...(2.18)}$$

The overall effectiveness of a screen can be expressed as

$$E = E_A \cdot E_B \qquad ...(2.19)$$

Substituting equation (2.17) and equation (2.18) in equation (2.19), we get

$$E = \frac{D \cdot x_D}{F \cdot x_F} \times \frac{B(1-x_B)}{F(1-x_F)} \qquad ...(2.20)$$

$$\Rightarrow \quad E = \frac{D}{F} \times \frac{x_D}{x_F} \times \frac{B}{F} \times \frac{(1-x_B)}{(1-x_F)} \qquad ...(2.21)$$

Substituting equation (2.14) and equation (2.16) in equation (2.21), we get

$$E = \frac{(x_F - x_B)}{(x_D - x_B)} \times \frac{x_D}{x_F} \times \frac{(x_D - x_F)}{(x_D - x_B)} \times \frac{(1-x_B)}{(1-x_F)}$$

$$E = \frac{x_D(x_F - x_B)(x_D - x_F)(1-x_B)}{(x_D - x_B)^2 x_F (1-x_F)} \qquad ...(2.22)$$

where, E = Screen effectiveness

x_D = Mass fraction of material A in overflow

x_B = Mass fraction of material A in underflow

x_F = Mass fraction of material A in feed

The effectiveness of a laboratory screen is as high as nearly unity, but for industrial screen is not so high. The screen effectiveness is also known as screen efficiency.

Example 2.1. An experimental analysis of a sample is made and the following data's are obtained.

$$x_F = 0.725$$

$$x_D = 0.850$$

$$x_B = 0.090$$

Calculate the screen effectiveness.

Solution:

$$E = \frac{x_D(x_F - x_B)(x_D - x_F)(1-x_B)}{(x_D - x_B)^2 x_F (1-x_F)}$$

From the given data, we have

$$x_F = 0.725$$

$$x_D = 0.850$$

$$x_B = 0.090$$

$$E = \frac{0.850 \times (0.725 - 0.090) \times (0.850 - 0.725) \times (1 - 0.090)}{(0.850 - 0.090)^2 \times 0.725 \times (1 - 0.725)}$$

$$= 0.5329$$

$$= 53.29\%$$

Ans.

Example 2.2. Experimental data of sieve analysis is given below. Calculate the screen effectiveness.

Assume, D_{PC} = 100 Mesh.

Sl. No.	Mesh No.	Aperture size (mm)	Mass retained		
			Feed (g)	Overflow (g)	Underflow (g)
1	50	0.297	16.3	10.0	—
2	72	0.211	34.3	29.5	—

Sl. No.	Mesh No.	Aperture size (mm)	Mass retained		
			Feed (g)	Overflow (g)	Underflow (g)
3	100	0.152	22.8	18.5	3.1
4	150	0.104	14.9	1.8	14.5
5	200	0.076	6.0	1.1	7.3
6	240	0.063	0.7	1.0	1.8
7	Pan	—	5.0	1.1	6.3
			100	63	33

Solution:

Sl. No.	Aperture size (mm)	Cumulative mass fraction retained		
		Feed	Overflow	Underflow
1	0.297	0.163	0.159	—
2	0.211	0.506	0.627	—
3	0.152	0.734	0.921	0.094
4	0.104	0.883	0.950	0.533
5	0.076	0.943	0.967	0.754
6	0.063	0.950	0.983	0.809
7	—	1.000	1.000	1.000

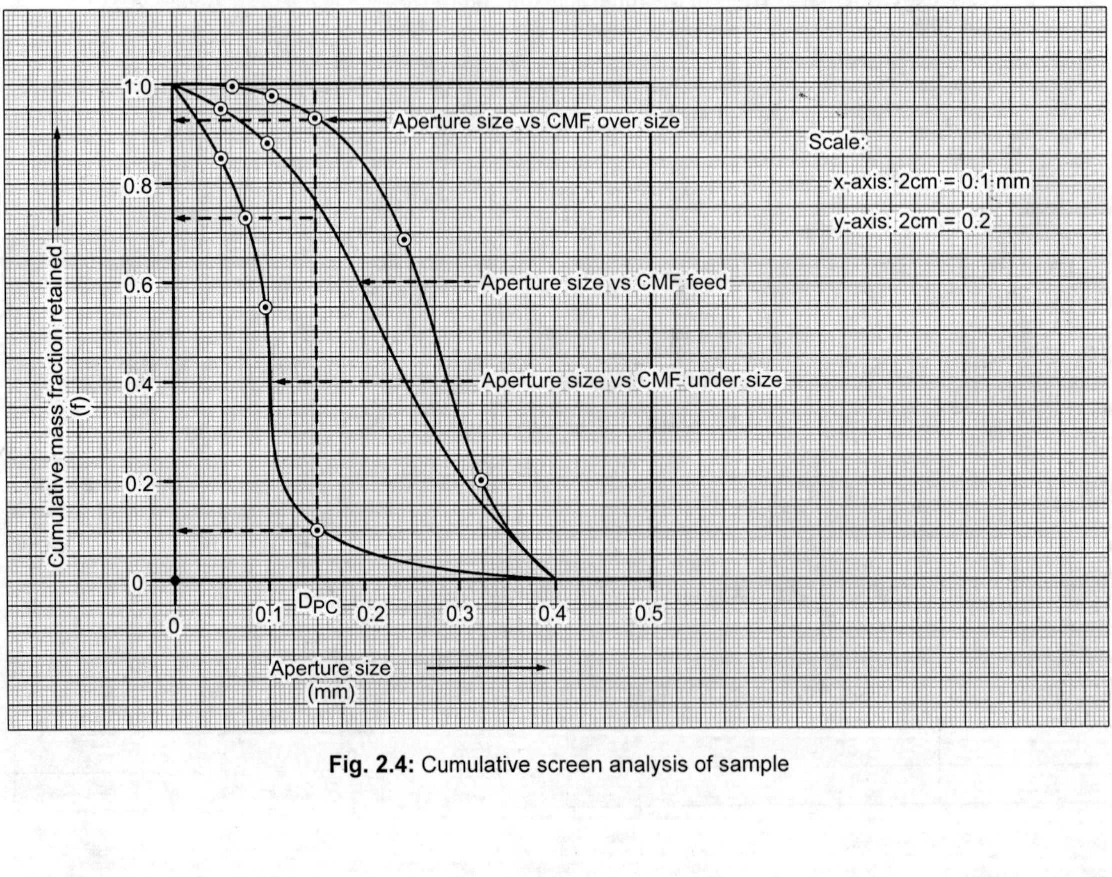

Fig. 2.4: Cumulative screen analysis of sample

From the graph, we have

$$x_F = 0.73$$
$$x_D = 0.92$$
$$x_B = 0.085$$
$$E = \frac{x_D(x_F - x_B)(x_D - x_F)(1 - x_B)}{(x_D - x_B)^2 x_F (1 - x_F)}$$
$$= \frac{0.92 \times (0.73 - 0.085) \times (0.92 - 0.73)(1 - 0.085)}{(0.92 - 0.085)^2 \times 0.73(1 - 0.73)}$$
$$= 0.7511$$
$$= 75.11\%$$

Ans.

Example 2.3. A quartz mixture having the screen analysis shown below is screened through 10 Mesh screen. Calculate the mass ratios of the overflow and underflow of feed and overall effectiveness of the screen.

Sl. No.	Mesh No.	D_P (mm)	Cumulative fraction larger than D_P		
			Feed	Overflow	Underflow
1	4	4.699	0.000	0.000	—
2	6	3.327	0.025	0.071	—
3	8	2.362	0.150	0.430	—
4	10	1.651	0.470	0.850	0.195
5	14	1.168	0.730	0.970	0.580
6	20	0.833	0.885	0.990	0.830
7	28	0.589	0.940	1.000	0.910
8	35	0.417	0.960	—	0.940
9	65	0.208	0.980	—	0.975
10	Pan	—	1.000	—	1.000

[VTU Exam Jan/Feb 2003]

Solution: From the graph, we have

$$x_F = 0.52$$
$$x_D = 0.92$$
$$x_B = 0.25$$

From the mass ratio of overflow to feed, we have

$$\frac{D}{F} = \frac{(x_F - x_B)}{(x_D - x_B)} \quad ...(2.16)$$

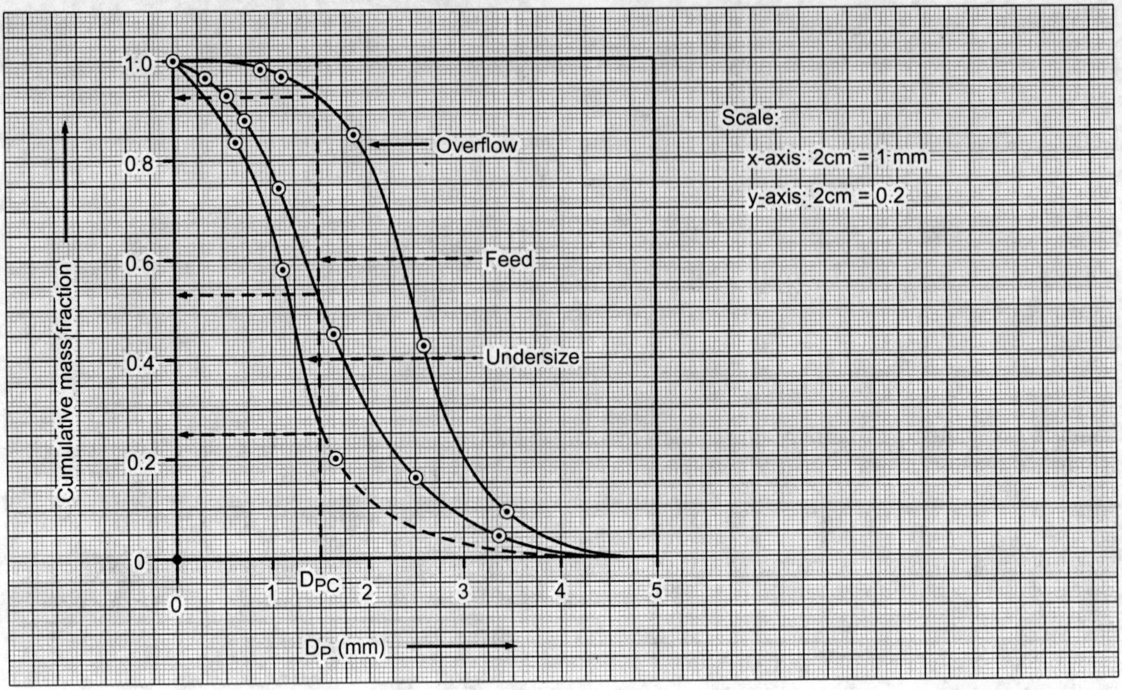

Fig. 2.5: Cumulative screen analysis

$$\Rightarrow \quad \frac{D}{F} = \frac{(0.52-0.25)}{(0.92-0.25)}$$

$$= 0.403 \qquad \textbf{Ans.}$$

From the mass ratio of underflow to feed, we have

$$\frac{B}{F} = \frac{(x_D - x_F)}{(x_D - x_B)} \qquad \text{...(2.14)}$$

$$\Rightarrow \quad \frac{B}{F} = \frac{(0.92-0.53)}{(0.92-0.25)}$$

$$= 0.597 \qquad \textbf{Ans.}$$

From the screen effectiveness, we have

$$E = \frac{x_D(x_F - x_B)(x_D - x_F)(1-x_B)}{(x_D - x_B)^2 x_F(1-x_F)}$$

$$= \frac{0.92 \times (0.52-0.25) \times (0.92-0.52) \times (1-0.25)}{(0.92-0.25)^2 \times 0.52 \times (1-0.52)}$$

$$= 0.6696$$

$$= 66.96\% \qquad \textbf{Ans.}$$

Example 2.4. From a crushed quartz sample the fraction less than 2 mm had to be removed by screening. The feed sample contained 35% of minus 2 mm material. After screening the oversize fraction contained 10% of minus 2 mm size and the undersize contained 82% of minus 2 mm size. Determine the efficiency of the screen.

Solution: From the given data, we have

$$x_F = 0.65$$
$$x_D = 0.90$$
$$x_B = 0.18$$

From the screen effectiveness, we haze

$$E = \frac{x_D(x_F - x_B)(x_D - x_F)(1 - x_B)}{(x_D - x_B)^2 x_F(1 - x_F)}$$

$$\Rightarrow \quad E = \frac{0.90 \times (0.65 - 0.18) \times (0.90 - 0.65) \times (1 - 0.18)}{(0.90 - 0.18)^2 \times 0.65 \times (1 - 0.65)}$$

$$\Rightarrow \quad E = 0.7351$$
$$= 73.51\%$$

Ans.

Example 2.5. Anthracite coal from a pulverization unit has been found to contain an excess of fine material 75% by weight. In order to remove these fines it is screened using 1.5 mm screen. Estimate the effectiveness of the screen using the following data:

Particle size (mm)	Mass fraction	
	Oversize from the screen	Undersize from the screen
– 3.33 + 2.36	0.143	0.000
– 2.36 + 1.65	0.211	0.098
– 1.65 + 1.17	0.230	0.234
– 1.17 + 0.83	0.186	0.277
– 0.83 + 0.59	0.196	0.149
– 0.59 + 0.42	0.034	0.101
– 0.42 + 0.29	0.000	0.141

[VTU Exam Aug/Sept 2000]

Solution:

Particle size (mm)	Cumulative mass fraction	
	Oversize from the screen	Undersize from the screen
2.845	0.143	0.000
2.005	0.354	0.098
1.410	0.584	0.332
1.000	0.770	0.609
0.710	0.966	0.758
0.505	1.000	0.859
0.355	1.000	1.000

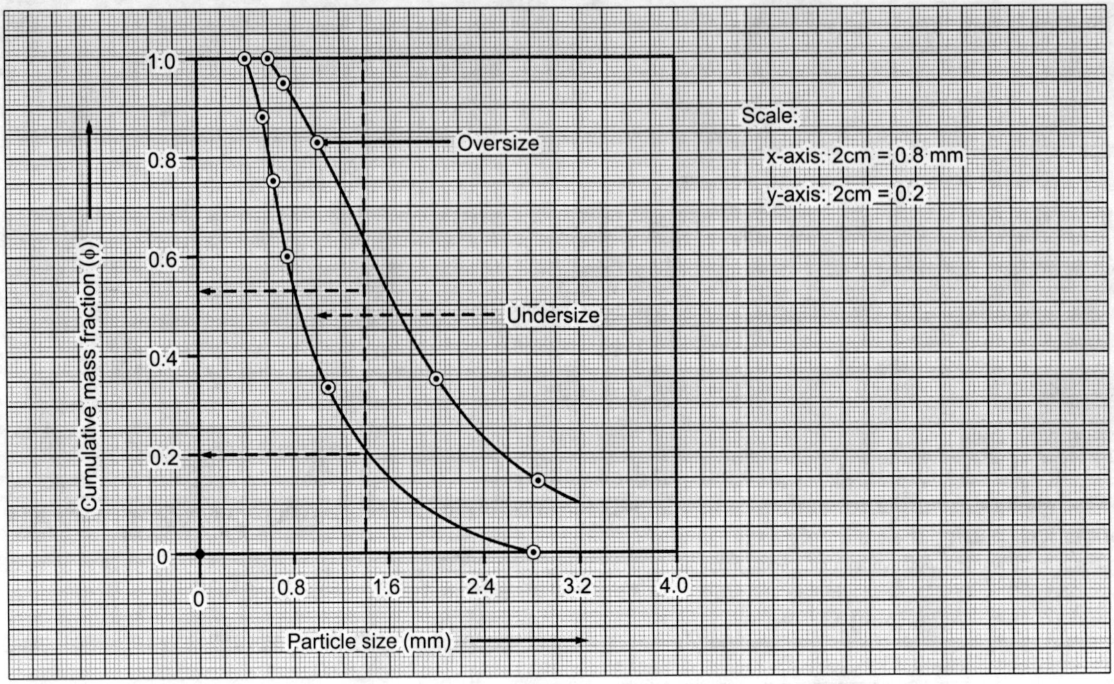

Fig. 2.6: Cumulative screen analysis

From the graph, we have

$$x_D = 0.54$$

$$x_B = 0.20$$

From the given data, we have

$$x_F = 0.25$$

From the screen effectiveness, we have

$$E = \frac{x_D(x_F - x_B)(x_D - x_F)(1 - x_B)}{(x_D - x_B)^2 x_F (1 - x_F)}$$

$$= \frac{0.54 \times (0.25 - 0.20) \times (0.54 - 0.25) \times (1 - 0.20)}{(0.54 - 0.20)^2 \times 0.25 \times (1 - 0.25)}$$

$$= 0.2890$$

$$= 28.90\%$$ **Ans.**

Screen effectiveness is 28.90%.

Example 2.6. In a laboratory experiment is carried out on screen analysis using stone particles. The following readings are recorded.

Sl. No.	Mesh No.	Weight retained (g)		
		Feed	Overflow	Underflow
1	30	23.1	23.1	0.0
2	52	16.2	16.1	0.0
3	72	14.3	14.5	0.0

4	100	10.4	10.8	0.0
5	120	0.8	0.6	0.1
6	150	11.1	9.0	0.2
7	170	6.9	0.3	9.1
8	200	6.7	0.0	7.5
9	Pan	10.5	0.0	8.7
		100.0	74.4	25.6

Determine the particle size distribution of the given sample and screen effectiveness.

Solution:

Sl. No.	Mesh No.	Aperture size, D_P (μm)	Weight fraction (x)			Cumulative weight fraction (ϕ)		
			Feed	Overflow	Underflow	Feed	Overflow	Underflow
1	30	500	0.231	0.310	0.0	0.231	0.310	0
2	52	300	0.162	0.218	0.0	0.393	0.528	0
3	72	212	0.143	0.192	0.0	0.536	0.720	0
4	100	150	0.104	0.145	0.0	0.640	0.865	0
5	120	125	0.008	0.008	0.004	0.648	0.873	0.004
6	150	106	0.111	0.121	0.008	0.759	0.994	0.012
7	170	90	0.069	0.004	0.355	0.828	1.000	0.367
8	200	75	0.067	0.0	0.293	0.895	1.000	0.660
9	Pan	—	0.105	0.0	0.340	1.000	1.000	1.000

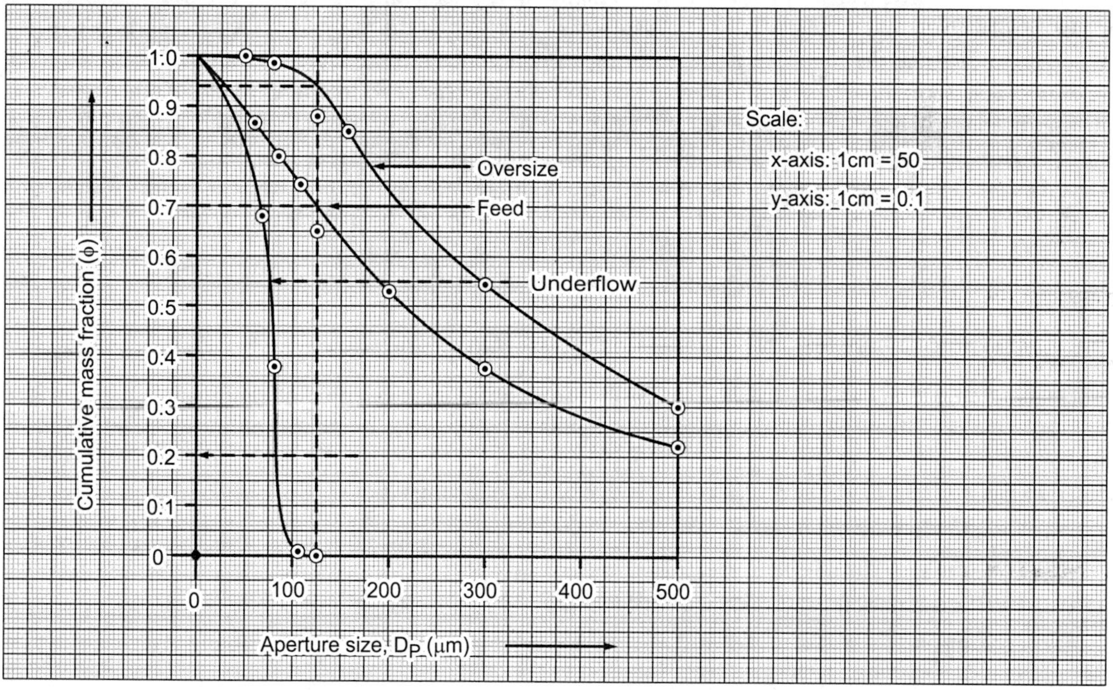

From the graph, we have

$$x_B = 0.01$$
$$x_F = 0.70$$
$$x_D = 0.945$$

From the screen effectiveness, we have

$$E = \frac{x_D(x_F - x_B)(x_D - x_F)(1 - x_B)}{(x_D - x_B)^2 x_F(1 - x_F)}$$

$$= \frac{0.945 \times (0.70 - 0.01) \times (0.945 - 0.70) \times (1 - 0.01)}{(0.945 - 0.01)^2 \times 0.70 \times (1 - 0.70)}$$

$$= 0.861$$

$$= 86.1\%$$ **Ans.**

Screen effectiveness is 86.1%.

2.12 CAPACITY AND EFFECTIVENESS OF SCREEN

The capacity may be defined as the measure of mass of material that can be fed per unit time to a unit area of the screen. These are the measure of performance of industrial screen. The capacity and effectiveness of screen operations are closely related. To obtain maximum effectiveness the screen capacity must be small and large capacity is attainable only at the expense of reduction in effectiveness of the screen. As the capacity and effectiveness are opposing factors, hence a reasonable balance between them is desired in actual practice. Accurate relationships are not available for estimating these characteristics of the screen but certain fundamental can be applied as guide in understanding the basic factors of the screen operations. The capacity of screen may be controlled by varying the rate of feed to the screening equipment, where as the effectiveness for a given capacity depends on the nature of the screening operation. The overall probability of passage of a given undersize particle of the sample is a function of the number of time the particle strikes's the screen surface and the chance of passage during a single contact. If the screen is overloaded, the number of contacts is less and the chance of passage on contact is reduced due to interference of the other particles. The improvement of effectiveness attained at the expense of reduced capacity of the screen. That is due to the more contacts per particle and better chances for passage on each contact. Ideally, a particle would have the greatest chance of passage through the screen if it struck the surface perpendicularly. The factors which tend to reduce capacity and lower effectiveness are binding and cohesion of the particle which act as barrier for the movement of particle to the screen surface. The moisture content of feed adversely affect the screen operation as it prone to stick the surface and each other.

2.13 MOTION OF SCREEN

In screening operation coarse particle drop easily through the large opening of stationary screen surface whereas in the case of fine particle do not pass through the screen easily. Therefore, for fine particle the screen must be agitated by shaking, gyrating or vibrating it mechanically or electrically. A typical screen motions are shown in Fig. 2.7. Motion of the screen could be one of the following type:

(*i*) Gyration in horizontal plane

(*ii*) Gyration in vertical plane

(*iii*) Gyration at one end and shaking at other end

(*iv*) Shaking

(*v*) Mechanically vibrated

(*vi*) Electrically vibrated

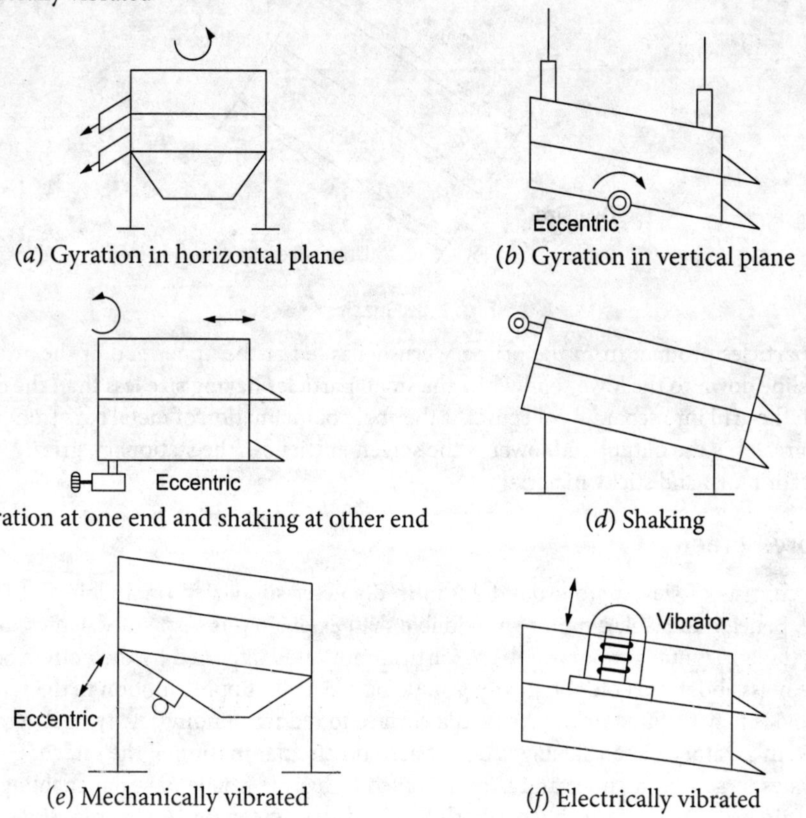

(*a*) Gyration in horizontal plane (*b*) Gyration in vertical plane
(*c*) Gyration at one end and shaking at other end (*d*) Shaking
(*e*) Mechanically vibrated (*f*) Electrically vibrated

Fig. 2.7: Motion of screen

2.14 SCREENING EQUIPMENT

Many varieties and types of screening equipment are available for different purposes, but only few are discussed here:

(*i*) Grizzly
(*ii*) Gyratory screen
(*iii*) Sieve shaker
(*iv*) Rotex screen
(*v*) Trommel screen
(*vi*) Riffled table
(*vii*) Wilfley table

2.14.1 Grizzly

A grizzly is a simple device consisting of a grid made up of metal bars usually built on a slope across which the sample material is passed. The scope and path of the sample material is parallel to the length of the metal bars. The length of the bar is around 3 m and spacing between the bar is 5 cm to 20 cm. The material of construction of the bar is Manganese steel, which helps to reduce the wear of the grizzly. Usually, the bar is constructed in such a way that its top is wider than the bottom. Hence, the bar can be made fairly deep for strength without being choked by the lumps of the sample passing through it. A typical grizzly is shown in Fig. 2.8.

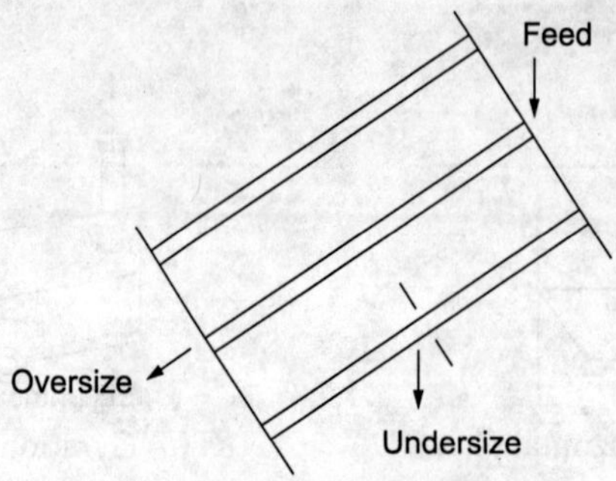

Fig. 2.8: Grizzly

The coarse particles product from the primary crusher is fed at the upper end of the grizzly. The large chunks roll and slide down to the lower end while the small particles having size less than the opening of the bars falls through the grid into separate collector. As the angle of inclination of metal bars from the horizontal plane is greater, greater is the output and lower is the screen efficiency. The stationary, grizzly is used for dry material, but not for moist and sticky material.

2.14.2 Gyratory Screen

Gyratory screen consists of sieve upto around 2.5 m in diameter supported on a table that is mounted on springs on a base. Beneath the table a motor with double shaft extensions is suspended, which drives eccentric weights to effect the horizontal gyratory motion. Vertical motion is imparted by the bottom weights, which swing the mobile mass about its center of gravity producing a circular tipping motion to the screen. Ball trays and ultrasonic device may be fitted below the screen surface to reduce blindings. A typical gyratory screen is shown in Fig. 2.9. In gyratory screen sieving is done based on circular motion of the screen. Unlike the other method, its operates in gentler manner and is more suited to handle fragile material enabling it to produce finer products. This method is applicable for both wet and dry screening. A distinct differences to other techniques is that the gyrator motion applied here depends on eccentric weight instead of vibration which can be varied based on individual process. These type of screens are typically used in low capacity applications. For example dewatering of carbon in gold plant.

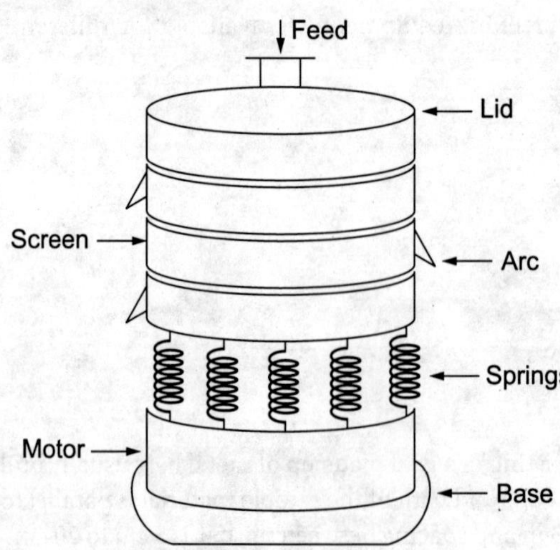

Fig. 2.9: Gyratory screen

The gyratory screen separation is also used by the pharmacy company for refining of powdary substances shaking away the excess material.

2.14.3 Sieve Shaker

It is a simple device used to determine the particle size distribution of a given sample. It consists of standard sieve set of different mesh numbers as per the specification of Indian Pharmacopoeia (IP) and United State Pharmacopeia (USP) and a mechanical shaker.

The materials of construction of the sieve are plenty, but in general Brass and Stainless Steel are mostly used. The sieves made of Brass are cheaper and Stainless Steel sieves are durable and versatile.

Sieves are arranged in a nest with coarsest one at the top and finest one at the bottom. A typical sieve shaker is shown in Fig. 2.10.

A sample is placed on the top sieve. This sieve set is fixed to the mechanical shaker, apparatus and shaken for a certain period of time minimum of 20 minutes. The particle sample retained on each screen is weighed.

Care should be taken inorder to get reproducible results. The type of motion influences sieving, vibratory motion is most efficient, followed by top hammering. This type of shaker is fixed and standard.

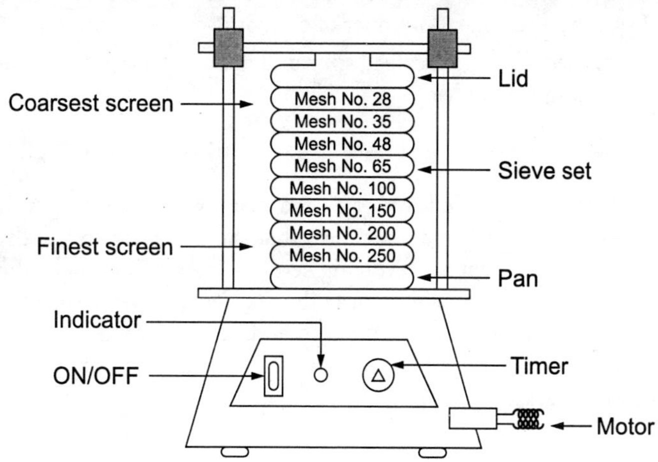

Fig. 2.10: Sieve shaker

It is simple, inexpensive and rapid with reproducible result. It has limitation of particle size less than 5 μm.

2.14.4 Rotex Screen

In Rotex screen instead of shaking or reciprocating motion, the screen is vibrated to keep the particle moving and to prevent the screen blinding. Hence, it is also known as vibrating screen. The vibrating may be accomplished by attaching to the screen pins that pass through the screen casing.

In this device the screens are nearly horizontal and underneath the screen much coarser supporting screens are attached. On these supporting screens at an intervals wooden blocks with sloping sides are placed. Between the two screens a number of rubber balls are provided. The shaking motion of the screen causes the rubber balls to fly about and strike the side of the wooden separating blocks. Hence, the inclined side of the wooden block deflect the balls upward against the undesirable of the screen. The whole screen mechanism is supported at lower end of the single bearing which is rotated by an eccentric pin on a heavy flywheel. A typical Rotex screen is shown in Fig. 2.11.

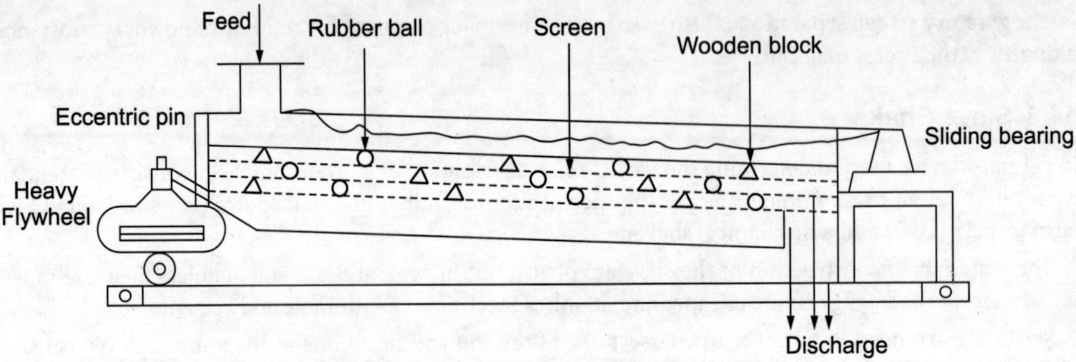

Fig. 2.11: Rotex screen

Vibrating screens can be made to perform one or more separations in a single piece of equipment. If there is one screen then only two products oversize and undersize are made. The coarser screens are placed at the top and finer at the bottom. Sufficient space is left in between the screens and each screen is made sufficiently shorter than the one above so that a discharge chute can be attached at the end of each screen. In such an operation if there are '*n*' number of screens, then '*n* + 1' number of discharge chute are provided to obtain the fraction.

Rotex screens are seldom used on a large scale in sizes under 4 mesh. For smaller particle separations they are frequently used upto 100 mesh.

2.14.5 Trommel Screen

It consists of a cylindrical frame surrounded by perforated plate which acts as screen. It is open at one or both end and slightly inclined to the horizontal plane so that the material can be advanced by the rotation of the cylinder. This device is also known as Revolving screen. The perforation of the screen may be same throughout or may be different at each section is near the feed end, then intermediate and the last coarse size. It is driven by a gear mechanism attached at the feed end. As the screen rotates, the undersize product discharge below the screen surface near the feed end and the oversize discharge at opposite end. A typical Trommel screen is shown in Fig. 2.12.

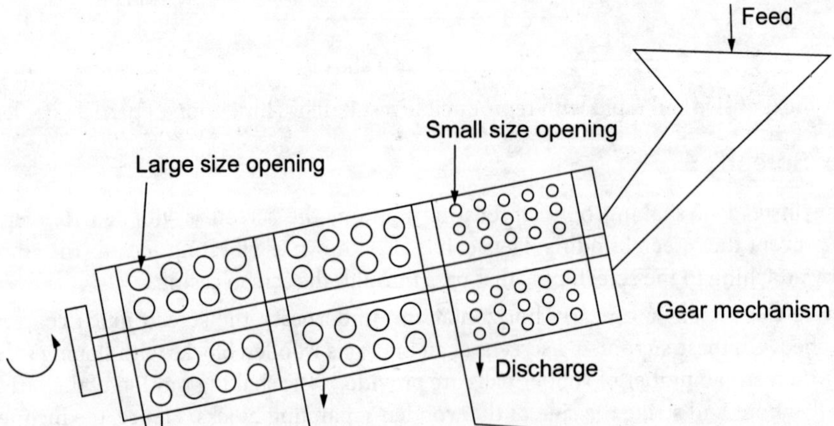

Fig. 2.12: Trommel screen

The material to be screened is fed at the upper end and gradually move down the screen surface towards the lower end. In doing so, the material passes over the apertures of gradually increasing size. If the single cylinder is provided with screen having three different size perforation then four fractions are obtained. The finest material is collected as underflow in the compartment near the feed end and the oversize material is withdrawn from the discharge end. This type of arrangement is usually adopted for smaller capacity screen.

In Trommel screen there is a tendency of blockage of the apertures by the large size particles. Also, the screen with finest opening being the weakest, it is subjected to the heaviest wear. The operating speed of the Trommel screen is 30 to 50% of the critical speed.

2.14.6 Riffled Table

It consists of a table provided with wooden riffles running approximately parallel to the top edge. The riffles are wooden strips of dimensions of about thickness 3 mm, width 9 mm and height 5 mm. They may be tapered from the feed end of the table to discharge end and are placed about 25 mm apart. The table is inclined at an angle 5° to the horizontal and is provided with adjustable tilting device so that the transverse slope of the table may be altered. A reciprocating motion is given to the table by the motor drive at a speed of 250 strokes per minute while handling with coarse material and 350 strokes per minute while dealing with fine materials. A typical riffled table is shown in Fig. 2.13.

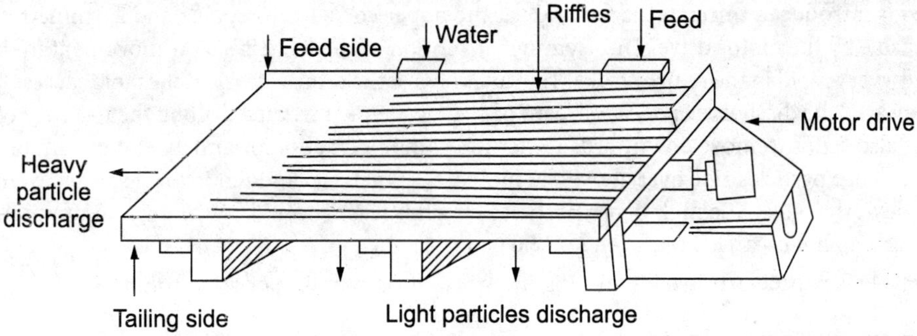

Fig. 2.13: Riffled table

The material containing solids of different densities is fed on to the top corner and a reciprocating motion is given to the table by the motor drive. The reciprocating motion of the table is a slow movement in the forward direction and very rapid is the return stroke. The quickness of return stroke of the table causes the materials to migrate towards the discharge end. The stream of water is introduced along the top edge of the table. The particles also tend to move downwards under the combined action of gravity and stream of water. The large and less dense particles ride over above the more dense material and tend to be washed over the riffles to the low side of the table. The high dense material which is least affected by the current of water, collect in the riffle along which it is carried to the end of the table.

Riffle table can be used to sort materials of different density from 7 mesh to 325 mesh.

2.14.7 Wilfley Table

A Wilfley table is almost similar to Riffled table. It consists of a table which has substantially a plane surface inclined slightly from the horizontal and shaken with an asymmetrical motion in the direction of long axis. The longitudinal tapered riffles usually made up of wood with a maximum height 10 mm and maximum width 10 mm are fixed on the table deck in the direction of table movement. The riffles are so placed that they form a channel of around 10 mm width and 10 mm deep which tapered to zero depth at the opposite end. The

2/3rd of the total surface area of the table is cleated and rest of the portion unriffled. The surface of the Wilfley table is lined with rubber or linoleum to restrict the wear of the wooden table surface and also to increase the roughness. An asymmetric motion is given to the table by the motor drive at a speed of 250 stroke per minute while dealing with coarse material and 350 stroke per minute while handling with fine material. A typical Wilfley table is shown in Fig. 2.14.

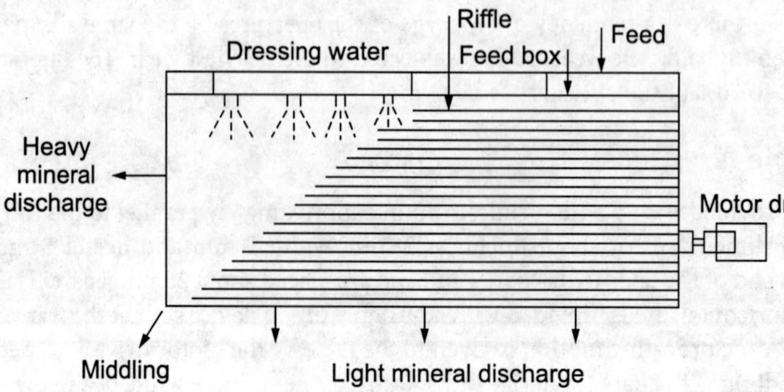

Fig. 2.14: Wilfley table

The feed is introduced through the feed box at the upper corner of the table and asymmetric motion is given to the table by the motor drive. The asymmetric motion of the table is a slow movement in the forward direction and a very rapid in the return stroke. The quickness of the return stroke of the table causes the material to migrate towards the discharge end. The stream of clean water is introduced along the top edge of the table. The particles also tends to move downwards under the combined action of gravity and stream of water. The large and less dense particles ride over above the more dense material and tends to be washed over the riffles to the low side of the table. The high dense particles which is least affected by the current of water, collect in the riffle along which it is carried to the end of the table.

Wilfley table is particularly used in mining industries for separation of gold from its ores.

2.15 SUB-SIEVE RANGE OF PARTICLES SIZE ANALYSIS

The particle which is small enough to pass freely through a 44 micron separating screen is called sub-sieve range of particles. The sub-sieve range of particles cannot be analyzed easily using a set of finer screen. Hence, they need to be analyzed by other methods. These other methods are called sub-sieve range of particles size analysis. Under this method the analysis can be carried out using the fluid media water and air.

Sub-sieve range of particle size analysis methods are plenty, but only a few are discussed here:

(*i*) Gravity settling tank

(*ii*) ICI sedimentation

(*iii*) Air elutriation

2.15.1 Gravity Settling Tank

It consists of a large settling tank with provision for suitable inlet and outlet. It works on the principle of settling rate of particles. Generally, the coarse particles have relatively faster settling rate than fine particles of the same specific gravity. Hence, the coarse particles will settled first, then intermediate and lastly the fines. This operation is also known as classifier. A typical gravity settling tank is shown in Fig. 2.15.

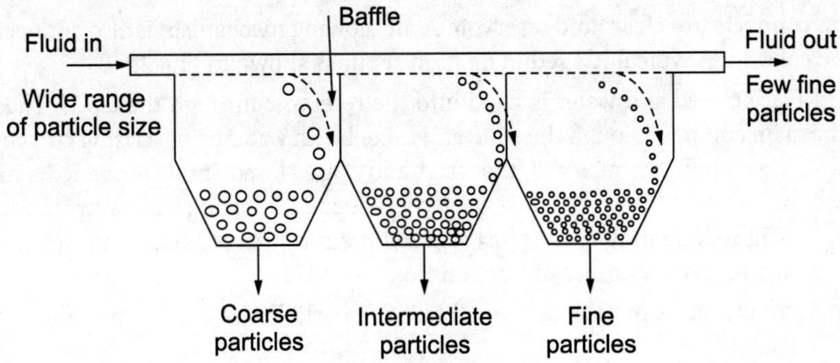

Fig. 2.15: Gravity settling tank

The slurry feed is allowed to enter the gravity settling tank through the inlet connection. As soon as the slurry feed enters the tank its linear velocity decreases due to enlargement of cross sectional area of the settling tank.

The particles start settling under the influence of gravity. The faster settling particles will be collected near the inlet while the slow settling particles will be carried further inside the tank before they reach the bottom of the tank. Placing vertical baffles within the tank allows for the collection of several fractions of the feed sample according to their terminal settling velocities. The very fine particles of the sample will be carried away along with the liquid overflow.

With this classifier the separation of the particles of the feed sample is not sharp, since considerable overlapping of size occurs.

2.15.2 ICI Sedimentation

Sedimentation is the tendency for particles in suspension to settle out of the fluid in which they are entrained and come to rest against a barrier. This is due to their motion through the fluid in response to the gravity and centrifugal forces acting on them.

The Imperial Chemical Industries (ICI) sedimentation column is generally made of glass taper at the bottom to a narrow tube a few centimetres long. It is fitted with taps at the top and bottom. A side branch in

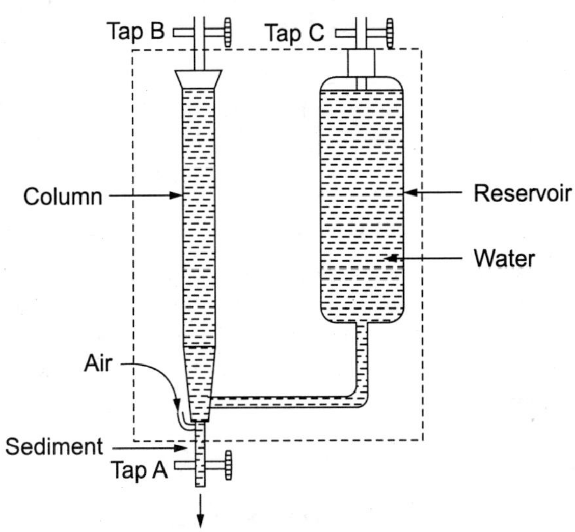

Fig. 2.16: ICI sedimentation

the narrow section connects to a clear fluid reservoir. A air blowing mechanism is also connects to the narrow tube section of the column. A typical ICI sedimentation set up is shown in Fig. 2.16.

The tap *A* and *B* are closed and water is filled into the reservoir through the tap *C*. Then, the sample is introduced into the main column through the tap *B* and mixed with water using a rising current of compressed air with tap *A* and *C* are closed. At time $t = 0$, the tap *A* and *B* are closed and *C* is opened, particles sediment into the narrow tube below the side branch. Now, the tap a is opened, allowing clear water to wash the sediment into a beaker. Neglecting sedimenting fluid is lost, the sediment is filtered, dried and weighed. The steps are repeated to obtained the required number of fractions.

ICI sedimentation technique may be used to obtain the particle size distribution for sub-sieve range of particles.

2.15.3 Air Elutriation

It is a process for separating particles based on their size, shape and density using a stream of air flowing in a direction usually opposite to the direction of sedimentation. This method is mainly used for particles smaller than 1 μm. The finer particles rise to the top because of their terminal sedimentation velocities are lower than the velocity of the rising fluid. The terminal velocity of any particles in any medium can be calculated using Stroke's law if the particles Reynold's number is less than 0.2.

An air elutriator is a simple device which can be used to separate particles into two or more fractions. It is consists of a vertical cylindrical tube made up of glass. It is fitted with air controlled mechanism and coarse particles outlet at the bottom of the column and a dome shape roof at the top with a provisions for air outlet and fines to collect. A typical air elutriator is shown in Fig. 2.17.

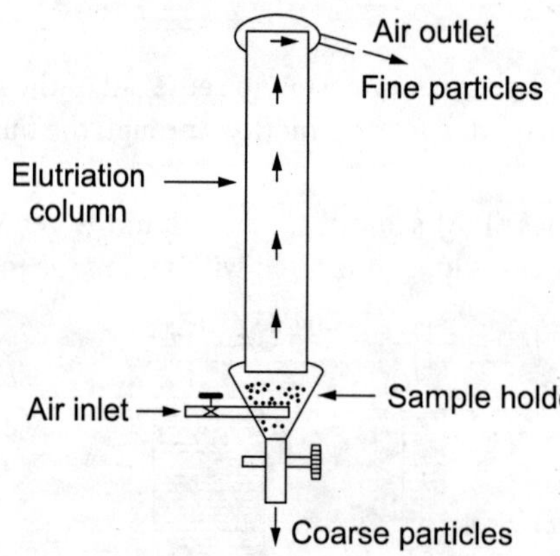

Fig. 2.17: Air elutriator

The sample is introduced into the sample holder and air is allowed to pass through it using air controlled mechanism. The smaller particles are carried over in the fluid stream while the larger particles settle against the upward current of air. If one starts with low flow rates, small low density particles attain their terminal velocities and flow with the stream. The particle from the stream is collected in overflow and hence will be separated from the feed sample. The flow rate can be increased to separate higher size ranges of particles.

2.16 SPECIFIC SURFACE AREA

It is the property of a solid material. Specific surface area may be defined as the total surface area of a material per unit mass of solid (m^2/kg) or bulk volume (m^2/m^3).

It is a physical value that can be used to determine the type and properties of a material. It has a particular importance for adsorption, heterogeneous catalysis and reactions on surfaces. The values obtained for specific surface area depend on the method of measurement. In adsorption based method the size of the adsorbate molecules, the exposed crystallographic planes at the surface and measurement temperature all affect the obtained specific surface area. For this reason, in addition to the most commonly used Brunaur Emmett Teller (BET) adsorption method, several techniques are available to measure the specific surface area of particulate material at ambient temperature. But, here only a few methods are discussed:

(*i*) Air permeability

(*ii*) BET adsorption method

2.16.1 Air Permeability

It may be defined as the rate of airflow passing perpendicularly through a known area under a prescribed differential air pressure between the two surface of the sample material.

The air permeability set up consists of a cylindrical tank made up of mild steel tapered at the bottom end to a tube a few centimetre long to which a tap is fitted. Also, a water level gauge is fixed vertically along the surface of the tank. Above the tank surface a pressure release valve and a sample bed chamber are fitted. The bed chamber consists of a fine screen at the bottom and a closure at the top. A differential manometer is attached between the two surface of the sample bed. A typical air permeability set up is shown in Fig. 2.18.

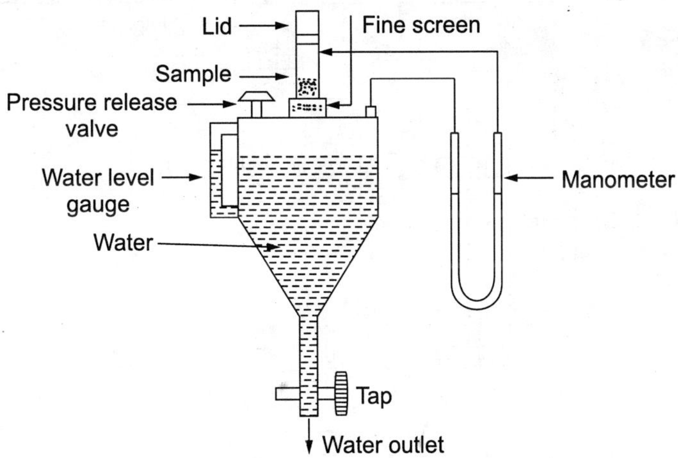

Fig. 2.18: Air permeability set-up

The 3/4th of the tank level is filled up with water. A known amount of sample around 8g is introduced into the sample bed. Then, the sample bed is tightly closed with a lid and fixed on the top flat surface of the tank. At $t = 0$, the water is allowed to run down through partially opened tap and collect in a beaker. This causes the suction of air through the bed porosity of the sample which create the differential pressure across the bed. The measurement of pressure drop, velocity and voidage provide and convenient way of measuring the specific surface area of particulate material. The specific surface area of the sample may calculated using the equation.

$$S_w = \sqrt{\frac{150 \cdot \Delta P \varepsilon^3}{(1-\varepsilon)^2 \rho_S^2 \bar{V} L H}} \quad ...(2.23)$$

where, S_w = Specific surface area, m^2/kg

ΔP = Pressure drop, N/m^2

ε = Porosity
ρ_s = Density of the sample, kg/m^3
$\bar{V}$ = Superficial velocity, m/s
L = Height of the bed, m
μ = Viscosity of air, kg/m.s

2.16.2 BET Adsorption Method

It works on the principle of physical adsorption of gas molecules on a solid surface and serves as the basis for an important analysis technique for the measurement of the specific surface area of materials. This method was developed by Stephen Brunauer, Paul Haugh Emmett and Edaward Teller in 1938 and published in the journal of American Chemical Society. This method may be applied for the system of multilayer adsorption which usually utilizes probing gases that do not chemically react with material surface as adsorbate to quantity specific surface area. In BET surface area analysis N_2 is usually used because of its availability in high purity and its strong interaction with most of the solids. For this reason, standard BET analysis most often carried out at the boiling temperature of N_2 (77K). Further probing adsorbate utilized in BET analysis are Argon, Carbon dioxide and water. A typical BET apparatus is shown in Fig. 2.19.

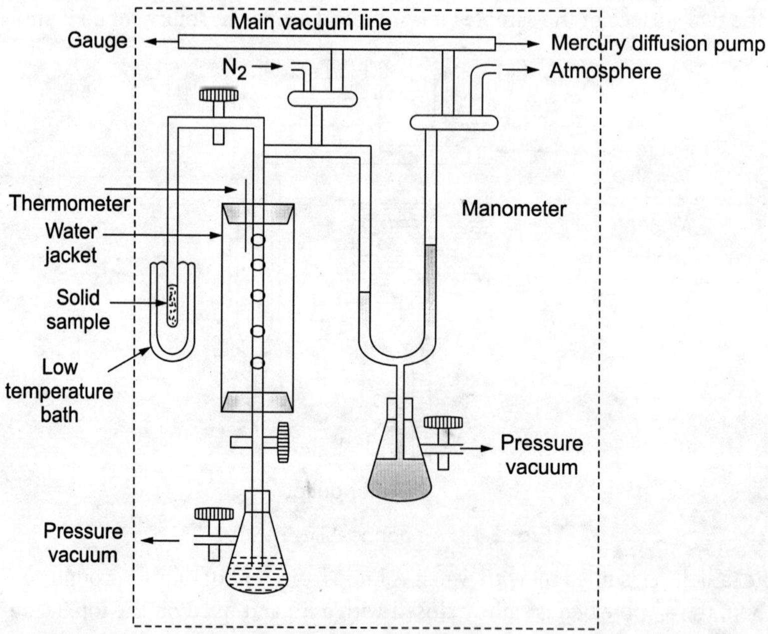

Fig. 2.19: BET apparatus

Specific surface area is a scale dependent property. Hence, it may be depend on the adsorbate molecules utilized and its adsorption cross-sectional area.

The BET adsorption method is commonly used to evaluate the gas adsorption data and generate a specific surface area. The result expressed in m^2/g. This method can be used to determine the specific surface area of porous material.

The equation using which specific surface area may be generated is

$$S_w = \frac{V_m Na}{22400m} \quad \text{...(2.24)}$$

where, S_w = Specific surface area, m^2/g
V_m = Volume of the gas adsorbed, ml
N = Avogaros constant, 6.023×10^{23} mol^{-1}
a = Effective cross-sectional area of one adsorbate molecule. For $N_2 = 0.162 \times 10^{-18}$ m^2
m = Mass of test powder, g

The volume occupied by one mole of adsorbate at STP is 22400 ml.

Example 2.6. Experiment is conducted on air permeability using stone powder as sample and following data are recorded. Calculate the specific surface area of the sample.

Data:

$$\varepsilon = 0.485$$
$$\Delta P = 361 \text{ N/m}^2$$
$$\rho_s = 2560 \text{ kg/m}^3$$
$$\bar{V} = 1.52 \times 10^{-2} \text{ m/s}$$
$$L = 0.029 \text{ m}$$
$$\mu = 1.93 \times 10^{-5} \text{ kg/ms}$$

Solution: From the air permeability equation, we have

$$S_w = \sqrt{\frac{150 \times \varepsilon^3 \times \Delta P}{(1-\varepsilon)^2 \times \rho_s^2 \times \bar{V} \times L \times \mu}}$$

$$\Rightarrow \quad S_w = \sqrt{\frac{150 \times (0.485)^3 \times 361}{(1-0.485)^2 \times (2560)^2 \times (1.52 \times 10^{-2}) \times (0.029) \times (1.93 \times 10^{-5})}}$$

$$= 646 \text{ m}^2/\text{kg}$$ **Ans.**

Example 2.7. The data generated by BET adsorption method are given below. Calculate the specific surface area of the sample. Assume N_2 is used as adsorbate.

Data:

$$V_m = 2.1 \text{ ml}$$
$$N = 6.023 \times 10^{23} \text{ mol}^{-1}$$
$$a = 0.162 \times 10^{-18} \text{ m}^2$$
$$m = 7\text{g}$$

Solution: From the BET adsorption method of analysis, we have

$$S_w = \frac{V_m N a}{22400 m} \quad \text{...(2.24)}$$

$$\Rightarrow \quad S_w = \frac{(2.1) \times (6.023 \times 10^{23}) \times (0.162 \times 10^{-18})}{22400 \times 7}$$

$$= 1.31 \text{ m}^2/\text{g}$$ **Ans.**

2.17 SPECIFIC SURFACE OF MIXTURE OF PARTICLE

If the particle density ρ_P and sphericity ϕ_S are known, then the surface area of the particles in each fraction may be calculated from the equation.

$$A = \frac{6m}{\phi_S D_P \rho_P} \quad \text{...(2.10)}$$

The results of all fractions are added to get S_w, the specific surface of a unit mass of particle. Then, S_w may be expressed as:

$$S_w = \frac{6x_1}{\phi_S D_{P_1} \rho_{P_1}} + \frac{6x_2}{\phi_S D_{P_2} \rho_{P_2}} + \ldots + \frac{6x_n}{\phi_S D_{P_n} \rho_{P_n}} \quad \ldots(2.25)$$

$$\Rightarrow \quad S_w = \frac{6}{\phi_S \rho_p} \sum_{i=1}^{n} \frac{x_i}{D_{P_i}} \quad \ldots(2.26)$$

where, S_w = Specific surface area of the particle sample, m^2/kg

ϕ_s = Sphericity

ρ_p = Density of the particle, kg/m^3

D_{p_i} = Average particle diameter, m

x_i = Mass fraction in a given increment

n = Number of increment

2.18 NUMBER OF PARTICLES IN A MIXTURE

The number of particles in a fraction is expressed as

$$N = \frac{m}{\rho_p V_p}$$

and the total population N_w in one mass unit of sample may be obtained by summation over the fraction as

$$N_w = \frac{x_1}{\rho_{p1} V_{p1}} + \frac{x_2}{\rho_{p2} V_{p2}} + \ldots + \frac{x_N}{\rho_{pn} V_{pn}} \quad \ldots(2.27)$$

For a given particle shape the volume of any particle is proportional to its diameter cubed.

$$V_p = aD_p^3 \quad \ldots(2.28)$$

Substituting equation (2.28) in equation (2.27), we get

$$N_w = \frac{x_1}{\rho_{p1} a D_p^3} + \frac{x_2}{\rho_{p2} a D_p^3} + \ldots + \frac{x_N}{\rho_{pn} a D_{pn}^3}$$

$$\Rightarrow \quad N_w = \frac{1}{a\rho_p} \sum_{i=1}^{n} \frac{x_i}{D_{P_i}^3}$$

$$\Rightarrow \quad N_w = \frac{1}{a\rho_p \bar{D}_V^3} \quad \ldots(2.29)$$

where, N_w = Number of particle in a mixture

a = Volume shape factor

ρ_p = Density of the particle, kg/m^3

$\bar{D}_V$ = Volume diameter of particle, m

REVIEW QUESTIONS

2.1 Define screening and state factors affect the screening operation.

2.2 List the operations used for separation of solid from solid and name one equipment of each category.

2.3 Define shape factor.

2.4 Discuss on particle size analysis.

2.5 What are the factors on which performance of screen mainly depends?

2.6 What is ideal screen and actual screen? Explain.

2.7 Define capacity of screen.

2.8 Define screen effectiveness.

2.9 Explain the term screen analysis.

2.10 Discuss briefly on differential screen analysis.

2.11 Discuss briefly on cumulative screen analysis.

2.12 Explain the term screen blinding.

2.13 How screen blinding is prevented? Explain.

2.14 What is standard sieve?

2.15 Name the standard screen used in engineering practice.

2.16 What are the advantages of expressing sieve by sieve number over the nominal size aperture?

2.17 What are the uses of screen analysis? How is it express?

2.18 Discuss on motion of screen.

2.19 With a neat diagram explain the construction and working of Grizzly.

2.20 With a neat diagram explain the construction and working of Gyratory screen.

2.21 With a neat diagram explain the construction and working of sieve shaker.

2.22 With a neat diagram explain the construction and working of Rotex screen.

2.23 With a neat diagram explain the construction and working of Trommel screen.

2.24 With a neat diagram explain the construction and working of Riffled table. Give its application.

2.25 With a neat diagram explain the construction and working of Wilfley table. Give its application.

2.26 Describe one industrial method for size separation of powder and its application.

2.27 What is sub-sieve range of particle size analysis?

2.28 With a neat diagram explain the construction and working gravity of settling tank.

2.29 With a neat diagram explain the construction and working of ICI sedimentation technique used for separation of sub-sieve range of particles.

2.30 With a neat diagram explain the construction and working of air elutriation technique used for the separation of sub-sieve range of particles.

2.31 With a neat diagram explain the construction and working of air permeability technique used for the measurement of specific surface area of particles.

2.32 With a neat diagram explain the construction and working of BET Adsorption method used for the measurement of specific surface area of particles.

2.33 Discuss the factors affecting the overall effectiveness of the screen.

2.34 Define mesh number.

2.35 Name few techniques used to measure the size of fine particles.

2.36 Define overall screen efficiency.

2.37 Give an account of the characterization of solid particles.

2.38 Discuss the interdependence of capacity and effectiveness in the case of screening.

2.39 Why cumulative analysis for determining the surface area is more precise than differential analysis?

2.40 Define volume surface mean diameter and arithmetic mean diameter of particles in a mixture.

2.41 Define mass mean diameter, volume mean diameter and number of particles in a mixture.

2.42 What is cut diameter in screening?

2.43 What is the relation between mesh size and capacity of the screen?

2.44 What are the distinctive properties of particulate masses?

2.45 Discuss the different methods of plotting screen analysis data and the specific uses of each method.

EXERCISES

2.1 Starting from the basic derive an expression for screen effectiveness.

2.2 Derive an expression for overall effectiveness of screen.

2.3 The screen analysis shown below applies to a sample of crushed quartz. The density of the particle is 2650 kg/m^3 and the shape factors are: a = 2 and f_S = 0.751. For the material between 4 mesh and 299 mesh in particle size calculate D_s and D_u.

Sl. No.	Mesh no.	Screen opening D_{pi}	Cumulative mass fraction X_i
1	4	4.699	0.0000
2	6	3.327	0.0251
3	8	2.362	0.1501
4	10	1.651	0.4787
5	14	1.168	0.7278
6	20	0.833	0.8868
7	28	0.589	0.9406
8	35	0.420	0.9616
9	48	0.297	0.9718
10	65	0.208	0.9795
11	100	0.147	0.9853
12	150	0.104	0.9894
13	200	0.074	0.9925
14	Pan	—	1.0000

2.4 A material is being fed to a screen 14 mesh size at the rate of 300 kg/hr. The desired product is material passing through the screen. The size analysis of the feed, coarse and fine streams are given below.

Mesh	Weight% obtained		
	Feed	Coarse	Fine
4	14.3	20.0	—
8	20.0	28.0	—
14	20.0	28.0	—
28	28.5	24.0	40.0
48	8.6	—	30.0
100	5.7	—	20.0
150	2.9	—	10.0

Calculate the effectiveness of screen and the rate of produce stream.

2.5 A material is feed to a double decker vibrating screen. The desired product is – 48 + 65 # fraction. The oversize analysis of the feed and three product streams are given below. Determine the effectiveness of the 48 mesh and 65 mesh screen.

Size	Feed	Overflow	Product	Underflow
– 20 + 28	9.0	19.0	1.6	0.0

– 28 + 35	18.0	39.0	4.0	0.1
– 35 + 48	26.0	33.0	32.4	0.6
– 48 + 65	28.0	6.0	53.0	34.3
– 65 + 100	9.0	1.5	6.3	30.0
– 100 + 150	6.0	1.5	2.5	24.0
– 150 + 200	4.0	0.0	0.2	11.0
	100	100	100	100

2.6 The sieve analysis is carried out for a sample powder with particle density 2536 kg/m^3. The data is given below. Assume sphericity is 0.9. Determine the following: Surface volume mean diameter and total surface area of the sample (m^2/g).

Mesh No.	60	80	100	115	150	170	200	250	270	325
Opening (mm)	0.25	0.177	0.149	0.125	0.105	0.088	0.074	0.063	0.053	0.044
Weight fraction % retained	0	5.6	10.8	14.3	23.4	16.2	12.9	8.3	5.3	3.2

2.7 From the experimental data given below, calculate the average particle size and surface area of the product both by differential and cumulative method of analysis. Given data's are ρ = 3.1 g/cm^3, $\phi_S = 0.571$.

Mesh no.	Clear opening (cm)	Weight retained (g)
7	0.2794	541
7/20	0.0853	139
20/36	0.0420	68
36/52	0.0300	60
52/72	0.0210	59
72/100	0.0150	43
100/150	0.0105	43
150/200	0.0075	21
200/350	0.0045	18
Pan	—	13

2.8. The following table gives the size distribution of a dust as measured by the microscope. Convert these figures to obtain the distribution on a weight basis and calculate the specific surface area, assuming spherical particles of specific gravity 2.65.

Size range (mm)	0-2	2-4	4-8	8-12	12-16	16-20	20-24
No. of particles in range	2000	600	140	40	15	5	2

[VTU Exam Nov. 2012]

2.9 In differential analysis of a quartz mixture, the volume surface mean diameter was determined as 1.25 mm. Calculate the specific surface area, if the density of quartz is 0.00265 g/mm^3. Shape factor can be taken as 0.62. **[Bangalore University Exam, April 2010]**

2.10 In a laboratory experiment is carried out on screen analysis using stone particles. The following readings are recorded.

Sl. No.	Mesh	Weight% obtained		
		Feed	Overflow	Underflow
1	30	34.00	33.85	0.0
2	52	17.80	16.96	0.0
3	72	20.07	20.09	0.0
4	100	11.35	11.22	0.0
5	120	4.55	3.81	1.09
6	150	2.05	0.0	1.56
7	170	3.24	0.0	3.78
8	200	2.57	0.0	2.75
9	Pan	4.37	0.0	4.95
		100.0	85.87	14.13

Determine the particle size distribution of the given sample and screen effectiveness.

❑❑❑

3 CHAPTER Size Reduction

3.1 INTRODUCTION

Size reduction is a process of reducing large solid particle into small solid particle or fine particles. It is also known as comminution or diminution or pulverization. Size reduction of bulk solids is a very common and important unit operation found across the process industries. Liberation of key species improved chemical activity or bioavailability and better handling characteristics are just a few of the reasons for using size reduction technology. Despite the availability of other particle production processes size reduction remains today one of the workhorses for production of bulk solids with tailored particles properties. This chapter focuses on the basic of how particles break individually and also in size reduction equipment. Since size reduction equipment rarely exists as a standalone entity the concept of grinding circuit will be explored with particular attention on the optimization of the whole circuit.

3.2 MECHANISM OF SIZE REDUCTION

The mechanism of size reduction may vary with the nature of material. Therefore, each material may required a separate treatment. However, the general mechanism of size reduction may be described as follows:

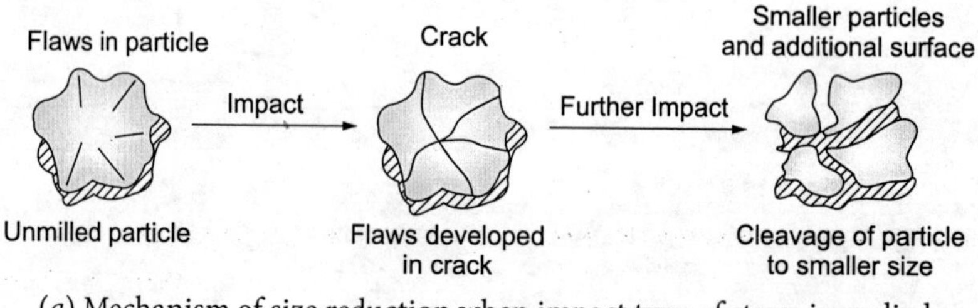

(*a*) Mechanism of size reduction when impact type of stress is applied.

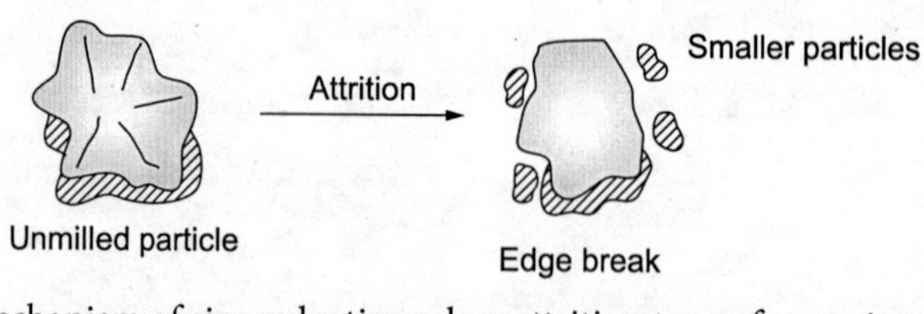

(*b*) Mechanism of size reduction when attrition type of stress is applied.

Fig. 3.1: Mechanism of size reduction

The particles whether it is amorphous or crystalline will have flaws to a definite degree. These flaws constitute weak part in the particles. When sufficient stresses such as impact, shear and compression are applied on the particles, the weak flaws develop into cracks which eventually leads to cleavage. Thus, smaller particles are obtained with additional surface area.

When stresses in the form of attrition is applied, the particle surfaces chip and produce small particles.

3.3 MODES OF FORCE APPLIED IN SIZE REDUCTION

In general forces of varied nature are required to achieve size reduction. The common modes of forces applied for size reduction are as follows:

(*i*) Cutting
(*ii*) Compression
(*iii*) Impact
(*iv*) Attrition

3.3.1 Cutting

The material is cut by means of a sharp blades. Example: Cutter mill.

3.3.2 Compression

In this mode the material is crushed between the roller by the application of pressure. Example: Roller mill.

3.3.3 Impact

In this mode involves the operation of hammer at high speeds. When a lump of material strikes the rotating hammers, the material splits apart. This action continues until particles of required size are obtained. Example: Hammer mill.

Impact also occurs when moving particles strike against a stationary surface. In the same way particles moving at high speeds collide each other and produce smaller particles. Example: Fluid energy mill.

3.3.4 Attrition

In this mode the process involves breaking down of the material by rubbing action in between two surfaces. Example: Fluid energy mill.

Although the predominant mode with one example of equipment is discussed, but normally more than one modes is exhibited for size reduction.

3.4 CRITERIA OF COMMUNICATION

Communication is a generic term for size reduction. Crusher and grinders are type of communicating equipment. An ideal crusher and grinder should

(*i*) have a large capacity
(*ii*) require a small power input per unit of product
(*iii*) yield a product of the single size or the size distribution desired.

The usual method of studying the performance of equipment is to set up an ideal operation as a standard compare to the characteristics of the actual equipment with those of the small unit and account for the difference between the two. When this method is applied to crushing and grinding equipment, the difference between the ideal and actual are very great despite expensive study the gaps have not been completely accounted for. On the other hand useful empirical equations for predicting equipment performance have been developed by the incomplete theory.

3.5 CHARACTERISTICS OF COMMINUTED PRODUCT

The objective of crushing and grinding is to produce small particles from larger ones. The small particles are desired either because of their more surface or because of their shape, size and number. One can measure the efficiency of operation based on the energy required to create new surface, since the surface area of unit mass of particles increases greatly as the particle size is reduced.

Like an ideal crusher or grinder an actual unit does not yield a uniform product whether feed is uniformly sized or not. The product always consists of a mixture of particles varying from a definite maximum to a very small particles. Some equipment designed to control the magnitude of the largest particles in their products, but the fine sized are not under control. In some type of crushers, fines are minimized but they are not eliminated. If the tail is homogeneous, both in the shape of the particles and in chemical and physical structure, then the shape of the individual units in the products may be quite uniform. Otherwise the grains in the various sizes of a single product may differ considerably in shape.

The ratio of the diameter of the largest and smallest particles in a communicated product is in order of 10^4. Because of this extreme variation in sizes of the individual particles, relationships adequate for uniform sizes must be modified when such mixture is communicated.

3.6 ENERGY AND POWER REQUIREMENT FOR SIZE REDUCTION

The cost of power requirement is a major expenses of crushing and grinding. During size reduction, the solid particles are first distorted and strained, thus work required to strain them is stored temporarily in solids as mechanical energy of stress. By applying additional force, stressed particles are distorted beyond their ultimate strength and suddenly fracture into small pieces. Thus, new surface area of the particle is generated. As the unit area of solid particle has a definite amount of surface energy, hence for the generation of new surface need to do work which is given by the release of energy of stress when the particles break. The energy of stress in excess of the new surface energy created appears as heat.

It is not possible to estimate accurately the energy and power requirement for size reduction of a given material by crushing and grinding. But, number of empirical laws have been put forward for the same purpose.

3.7 LAWS OF SIZE REDUCTION

The laws governing size reduction are as follows:

(*i*) Rittinger's law

(*ii*) Kick's law

(*iii*) Bond's law

3.7.1 Rittinger's Law

According to the Rittinger's law energy required for the size reduction of unit mass of particle is directly proportional to the new surface created. Mathematically, the law can be expressed as:

$$\frac{P}{\dot{m}} = K_r\left[\frac{1}{\bar{D}_{sb}} - \frac{1}{\bar{D}_{sa}}\right] \qquad ...(3.1)$$

where, P = Power required, kW

$\dot{m}$ = Feed rate, t/hr

K_r = Rittinger's law constant

$\bar{D}_{sa}$ = Volumetric surface mean diameter of feed, m

$\bar{D}_{sb}$ = Volumetric surface mean diameter of product, m

This law is found to suitable for the cases where the energy input per unit mass of solid particle is not too high.

3.7.2 Kick's Law

According to the Kick's law energy required for the size reduction of unit mass of particle is directly proportional to the logarithm of the ratio between the initial and final diameter of the solid particle. Mathematically, the law can be expressed as:

$$\frac{P}{\dot{m}} = K_k \ln \frac{d_1}{d_2} \qquad ...(3.2)$$

where,
P = Power required, kW
$\dot{m}$ = Feed rate, t/hr
K_k = Kick's law constant
d_1 = Initial diameter of solid particle, m
d_2 = Final diameter of solid particle, m

As the energy required for size reduction is directly related to the reduction ratio between the initial and final diameter of the solid particle, hence the energy required to crush a given quality of material from 60 mm to 30 mm is the same as that required to reduce the size from 30 mm to 15 mm.

Kick's law is found to be more accurate than Rittinger's law for coarse particle crushing where the amount of surface produced is considerably less.

3.7.3 Bond's Law

According to the Bond's law energy required to form a particle size D_p from a very large feed is directly proportional to the square root of the surface to volume ratio of the product. Mathematically, the law can be expressed as:

$$\frac{P}{\dot{m}} = \frac{K_b}{\sqrt{D_p}} \qquad ....(3.3)$$

where,
P = Power required, kW
$\dot{m}$ = Feed rate, t/hr
K_b = Bond's law constant
D_p = Particle size, mm

Bond's law is suggested for intermediate between the Rittinger's law and Kick's law for estimating the power requirement for size reduction by crushing and grinding operation.

Bond's law is somewhat more realistic in estimating power requirements of commercial size reduction equipment.

For using equation (3.3), a work index W_i may be defined as the amount of gross energy is kilowatt-hours per ton of feed material required to reduce a very large feed to such size that 80% of the product passes through a 100 μm screen.

If D_p is in millimeter, P in kilowatts and $\dot{m}$ in tons per hour, then the relation between K_b and W_i, based on the definition of work index may be expressed as

$$K_b = \sqrt{100 \times 10^{-3}} W_i \qquad ...(3.4)$$

$$K_b = 0.3162\ W_i \qquad ...(3.5)$$

Substituting equation (3.5) in equation (3.3), we get

$$\frac{P}{\dot{m}} = 0.3162W_i\left[\frac{1}{\sqrt{D_{pb}}}\right] \quad ...(3.6)$$

If 80% of feed passes a mesh size of D_{pa} mm and 80% of the product passes a mesh size of D_{pb} mm, then the equation (3.6) may be explored as

$$\frac{P}{\dot{m}} = 0.3162W_i\left[\frac{1}{\sqrt{D_{pb}}} - \frac{1}{\sqrt{D_{pa}}}\right] \quad ...(3.7)$$

where,

W_i = Work index

D_{pa} = Particle diameter of feed, mm

D_{pb} = Particle diameter of product, mm

For large particle size of feed the term $1/\sqrt{D_{pa}}$ of equation (3.7) may be neglected.

Work index include the friction in the crusher. Therefore, power estimated by the equation (3.7) is gross power. Work index of some commonly used materials are given in Table 3.1.

Table 3.1: Work index of materials

Sl. No.	Materials	Work Index (W_i) kWh/t
1	Clay	6.30
2	Gypsum rock	7.40
3	Phosphate rock	10.50
4	Limestone	12.65
5	Coal	11.34
6	Quartz	13.60
7	Gravel	16.00

3.8 ENERGY UTILIZATION

Size reduction begins with the opening of small flaws that were usually present. Based on the probability, larger particles may have numerous flaws and easily fractured when compared to smaller particles with fewer flaws. The supplied energy used to increase new surface area from flaws and also to initiate flaw or cracks in the particles. This phenomena is observed generally in fine grinding. Most efficient mills utilizes less than 1% of the energy input to fracture particles or create new surfaces. Rest of the energy is dissipated in

(*i*) elastic deformation or unfractured particles.

(*ii*) transport of material within the milling chamber

(*iii*) friction between particles.

(*iv*) heat of friction between particles and mill.

(*v*) vibrating and noise in effecting of transmission.

Energy input and utilization are important factors to estimate the efficiency of mill for a given material. A number of theories have been proposed to establish a relationship between energy and the degree of size reduction produced. Energy required to reduce the size of the particle is inversely proportional to the size D to some power. This may be mathematically expressed as

$$\frac{dE}{dD} = \frac{C}{D^n} \quad ...(3.8)$$

where, E = Energy required to produce a change, kwh
D = Size of the unit mass, mm
C = Constant
n = Number

At $n = 1$, integrating the equation (3.8) between the limit, we get

$$E = C\ln\frac{d_1}{d_2} \quad ...(3.9)$$

where, d_1 = Diameter of feed particle, mm
d_2 = Diameter of product particle, mm

In equation (3.9), the ratio of d_1/d_2 is known as reduction ratio. The usefulness of equation (3.8) is that it satisfied various laws proposed. For example:

For $n = 1$, equation (3.8) gives the solution of Kick's law

For $n = 1.5$, equation (3.8) gives the solution of Bond's law

For $n = 2$, equation (3.8) gives the solution of Rittinger's law

3.9 CRUSHING EFFICIENCY

It may be defined as the ratio of the surface energy created by crushing to the energy absorbed by solid. Mathematically, it may expressed as

$$\eta_c = \frac{e_s(A_b - A_a)}{W_n} \quad ...(3.10)$$

where, η_c = Crushing efficiency
e_s = Surface energy per unit area, J/m^2
A_b = Area of the product, m^2
A_a = Area of the feed, m^2
W_n = Energy absorbed by the material, J/kg

The energy created by fracture is very small as compared to the energy stored at the time of rupture of the particle, hence most of the mechanical energy stored in material is converted into heat. Therefore, crushing efficiencies are low.

The energy absorbed by the solid is less than energy supplied to the crusher is part of the total energy, supplied to the machine is utilized to overcome the friction in the bearings and other moving parts and the remaining is available for crushing.

The mechanical efficiency may be defined as the ratio of the energy absorbed to the energy input. Mathematically, it may be expressed as

$$\eta_m = \frac{W_n}{W} \quad ...(3.11)$$

where, η_m = Mechanical efficiency
W_n = Energy absorbed by the solid
W = Energy input to the crusher

$$\Rightarrow \quad W = \frac{W_n}{\eta_m} \quad ...(3.12)$$

Substituting equation (3.10) in equation (3.12), we get

$$W = \frac{e_s(A_b - A_a)}{\eta_m \eta_c} \quad ...(3.13)$$

where, A_a = Specific surface area of feed

A_b = Specific surface area of the product

The specific surface area of the particles may be calculated using the equation

$$A = \frac{6}{\phi_s \rho_p \bar{D}_s} \quad ...(3.14)$$

where, A = Specific surface area

ϕ_s = Sphericity

ρ_p = Particle density

$\bar{D}_s$ = Volume surface mean diameter

3.10 METHOD OF OPERATION OF CRUSHER AND GRINDER

The crusher and grinder may be operated under following condition:

(*i*) Free crushing

(*ii*) Choke crushing

(*iii*) Open circuit and closed circuit grinding

(*iv*) Wet and dry grinding

3.10.1 Free Crushing

Whether crushing and grinding follows Rittinger's or Kick's law, the power is largely consumed in crushing by the finer particles. Consequently, if it is desired to reduce a given material say 10 mesh, it is obviously a waste of power if any appreciable portion of the product is say 50 mesh. To ensure a minimum of undersized material in the product, it is generally desirable that the feed to any crushing device to be so limited that the product drops freely away from the crushing surface and there is no undue clogging of the surface or packing of fine material around the coarser lumps. This is known as free crushing.

3.10.2 Choke Crushing

To ensure a minimum of undersized material in the product. It is generally desirable that the feed to any crushing device to be so limited that the product does not drops freely away from the crushing surface and there is no undue clogging of the surface or packing of fine material around the coarser lumps. This is known as choke crushing.

Many machine such as the jaw crusher, crushing roll and similar machine, will be free themselves of the products if not fed too heavily and the material handled is not damp or sticky materials some accessory devices may be required to prevent choke feeding. Various automatic feeding have been developed, applicable to many types of machines. A grinding machine has been operated with a stream of water or solution flowing over the material to be crushed. The purpose of this is to carry away the fine particles and leave only the coarse one for the action of crushing surface. This is also the reason for the air separation method used in the Raymond and similar mills. Still another method for preventing choke feeding, for reducing water consumption and preventing of excessive formation of undersize material.

3.10.3 Open Circuit and Closed Circuit Grinding

The size reduction process may be achieved in open circuit or closed circuit conditions. An open circuit mill is one in which milling is carried out in one attempt i.e., by passing the feed material through the mill to obtain the desired size.

A closed circuit mill is one in which the discharge from the milling process is passed through the size separation device or classifier and the oversize particles are returned to the grinding chamber for further size reduction. A typical closed circuit mill is shown in Fig. 3.2.

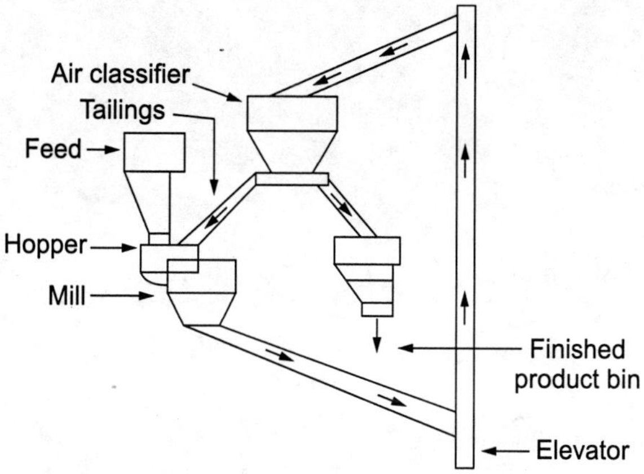

Fig. 3.2: Closed circuit grinding

The principle of closed circuit grinding is widely used in the pharmaceutical industry. It has the following advantages:

(*i*) It provides fine and ultrafine

(*ii*) It prevents dust

(*iii*) It eliminates overheating

(*iv*) It produces a product that is free from oversize or undersize.

In this type, size reduction is combined with size classification. The coarse size material returning to the mill by a classifier is known as circulating load.

3.10.4 Wet and Dry Grinding

The use of a product and subsequent steps mainly decide the type of equipment to be used. In the preparation of colloidal dispersions, emulsions and ointments, wet grinding offer advantages. In the production of suspensions and emulsions, wet grinding has become an integral part of the processing. Roller mill, colloid mill are used for this purpose.

In practice, it is found that finer size can be achieved by wet grinding than by dry grinding. Wet grinding is used in the production of flocculated suspensions. In wet grinding surfactants are added to attain particle size 0.3 mm. The advantages of wet grinding are as follows:

(*i*) It eliminates the dust hazards

(*ii*) Wet grinding is usually done at low speed, therefore it consumes low power.

(*iii*) Grinding aid can be incorporated

(*iv*) Colloid mill and ball mill are the best suited for wet grinding.

If the product undergoes physical or chemical changes in water, dry grinding is recommended. The fineness is limited to 100 μm. Hammer mill, roller mill, edge runner mill are used for dry grinding.

3.11 CLASSIFICATION OF SIZE REDUCTION EQUIPMENT

It is extremely difficult to make a rigid classification of size reduction equipment because of the wide variety of devices used. The only classification in which definite limitations of the groups can be established is the division into coarse crusher's, intermediate crusher and fine grinders. The different devices may be classified under these heads are as follows:

I. Coarse crushers:

(*i*) Jaw crushers
(*ii*) Gyratory crushers
(*iii*) Crushing rolls

II. Intermediate crushers:

(*i*) Hammer mill
(*ii*) Rolling compression mill
(*iii*) Attrition mill
(*iv*) Tumbling mill

III. Ultrafine grinders:

(*i*) Classifying hammer mill
(*ii*) Fluid energy mill
(*iii*) Colloidal mill
(*iv*) Agitator bead mill
(*v*) Edge runner mill
(*vi*) End runner mill

IV. Cutting machines:

(*i*) Rotary knife cutter mill

3.11.1 Coarse Crushers

Coarse crushers are the machine used for crushing big size particle into small pieces. The mainly used coarse crusher machine in laboratory and small scale industries are:

(*i*) Jaw crushers
(*ii*) Gyratory crushers
(*iii*) Crushing rolls

3.11.1.1 Jaw Crusher

Jaw crushers are two type

(*i*) Blake jaw crusher
(*ii*) Dodge jaw crusher

Blake jaw crusher: American inventor Eli Whitney Blake best known for his stone crushing machine. He has invented the Jaw crusher in 1851.

The working principle of Blake jaw crusher is based on the reciprocating movement of the movable jaw that compresses and crushes the material between itself and the fixed jaw.

The jaw crusher consists of a set of vertical jaws one jaw being fixed and the other being moved back and forth relative to it by a cam or pitman mechanism. The jaws are further apart at the top than at the bottom, forming a tapered chute so that the material is crushed progressively smaller and smaller as it travels downwards until it is small enough to escape from the bottom opening. A typical Blake jaw crusher is shown in Fig. 3.3.

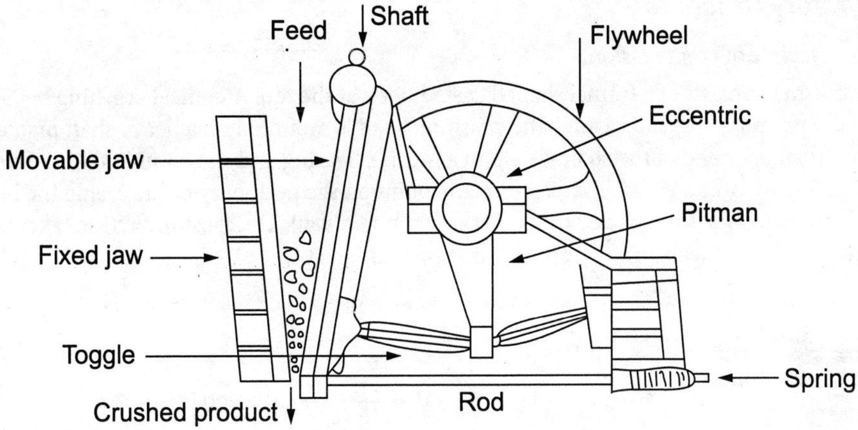

Fig. 3.3: Blake jaw crusher

The crushing angle in standard Blake jaw crusher is 27°. The reduction ratios at minimum recommended settings and with straight jaw plates average is about 8:1. Curved jaw plates are designed to minimize choking. The overhead eccentric jaw crusher falls into the second type. These are single toggle machines.

The main advantage of Blake jaw crusher is simple structure and easy maintenance.

Dodge jaw crusher: Dodge jaw crushers are the crushers wherein the moving plate is pivoted at the bottom and connected to an eccentric shaft. In universal crushers the plates are pivoted in the middle so that both the top and the bottom ends can move. The movable jaw is hinged at the bottom of the crusher frame so that the maximum amplitude of motion is obtained at the top of the crushing jaws. A typical Dodge jaw crusher is shown in Fig. 3.4.

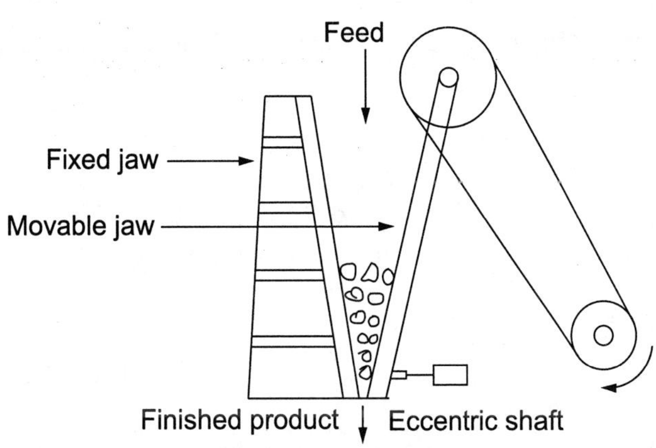

Fig. 3.4: Dodge jaw crusher

They are comparatively lower in capacity than the Blake jaw crushers and are more commonly used in laboratories.

In Blake jaw crusher the movable jaw is pivoted at the top. Therefore, the greatest amount of motion is at the bottom which means it has the little tendency to choke.

In Dodge jaw crusher the moving jaw is pivoted at the bottom. As minimum movement is at the bottom it has a greater tendency to choke.

3.11.1.2 Gyratory Crushers

It work on the principle of compression.

Gyratory crusher consists of a funnel shaped casing open at the top. A conical crushing head in the form of a truncated cone gyrates inside a casing. The crushing head is mounted on a heavy shaft pivoted at the top of the machine. The upper end of the shaft is held in a flexible bearing and the lower end of the shaft is driven by an eccentric 30 as to force a circle. Therefore, at any point on the periphery of the casing the bottom of the crushing head moves forward and backward from the stationary wall. The crushing action takes place around the whole of the cone. A typical gyratory crusher is shown in Fig. 3.5.

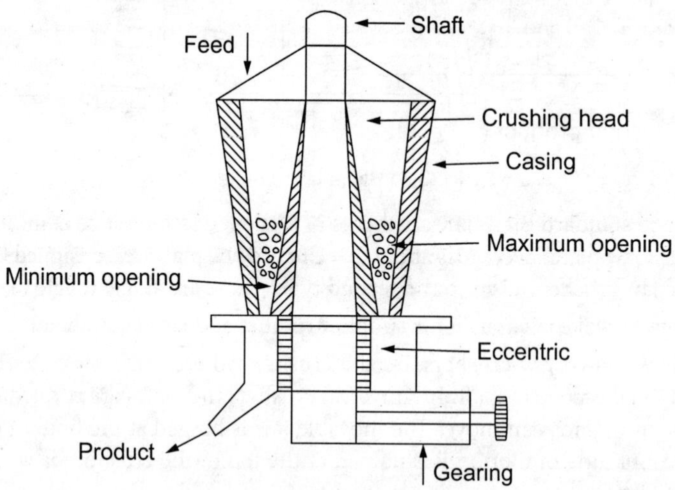

Fig. 3.5: Gyratory crusher

The material to be crushed is charged from the top of the machine. The conical head gyrates in the casing. At any point on the periphery of the casing, the bottom of the crushing head moves towards and then away from the stationary wall. The solids caught in the V-shape space between the head and the casing are broken and rebroken until they drop out from the bottom of the machine. The speed of the crushing head usually lies between 120 to 480 gyration per minute. Since some part of the crushing head is working at all times; hence the discharge from this crusher is continuous instead of intermittent as in a Blake jaw crusher.

Gyratory crushers are used for initial size reduction to convert big rocks into small stone.

3.11.1.3 Crushing Roll

In this crusher the material is crushed by the application of stress, though attrition also influences. Stress is applied by rotating heavy wheel, muller or roller.

The crushing roll consists of two cylindrical roller made of stone or metal which are mounted horizontally.

The roller may have diameter about one metre. Rollers are capable of rotating on their longitudinal axes. Generally, one of the roller is driven directly using a motor, while the other one rotate freely. The gap between the roller can be controlled to obtain the desired particle size. A typical crushing roll is shown in Fig. 3.6.

The roller are allowed to rotate. The material is fed from the hopper into the gap between the two rollers. The material is crushed while passing through the roller under high pressure. The gap between the rollers can be adjusted to control the degree of size reduction. The product is collected into a receiver.

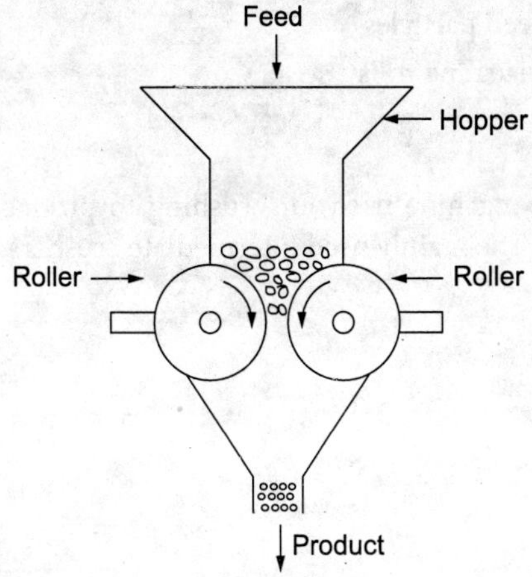

Fig. 3.6: Crushing roll

Crushing rolls are effectively used in minerals crushing where the ores are not too abrasive and they are also used in small and scale production mining of more abrasive metal ores such as gold.

Angle of nip in roll crusher: The nip angle is defined as the angle that is tangent to the roll surface at the point of contact between the rolls and the particles. It depends on the surface characteristics of the rolls. Usually, the nip angle is between 20° and 30°. But in some large roll crushers it is upto 40°. Consider the two rolls crusher as shown in Fig. 3.7.

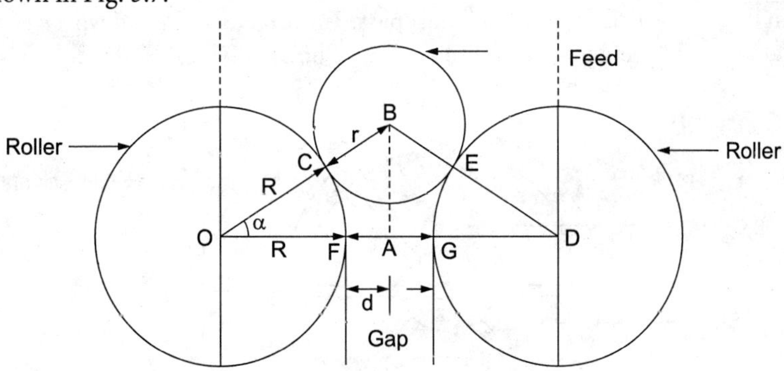

Fig. 3.7: Angle of nip in Rolls crusher

Consider the triangle OAB:

From the basic trigonometry, we have

$$\cos \alpha = \frac{OA}{OB}$$

$$\Rightarrow \qquad \cos \alpha = \frac{OF + FA}{OC + CB}$$

$$\Rightarrow \qquad \cos\alpha = \frac{R+d}{R+r}$$

$$\Rightarrow \qquad \cos\alpha = \frac{d+R}{r+R} \qquad ...(3.15)$$

where, α = Angle of nip

d = Gap between the rolls/2

r = Radius of the feed particles

R = Radius of the crushing rolls

3.11.2 Intermediate Crushers

The intermediate crushers are the machine used for crushing to produce various coarser fractions or to prepare material for final crushing. The mainly used intermediate crushers are as follows:

(*i*) Hammer mill

(*ii*) Rolling compression mill

(*iii*) Rolling compression mill

(*iv*) Attrition mill

(*v*) Tumbling mill

3.11.2.1 Hammer Mill

The hammers mill operates on the principle of impact between rapidly moving hammers mounted on a rotor and the feed materials.

The hammer mill can be either horizontal or vertical shaft type. Hammers are usually made up of hardened steel with impact surface made of an extremely abrasive resistant material such as Haystellite and Carbaloy. The basic shape of the hammers are stirrup and bar. The hammer blade can be with flat edge or sharp edge or both on each side. It may be either rigid or swing type. Free swinging type has an advantage that there will be increasing clearance between hammers and screen. This unit is enclosed with a chamber consisting of a grid or removable screen through which the material must pass. The screens are not of woven type. the screens are prepared using metal sheet of varying thickness with perforated holes. A typical Hammer mill is shown in Fig. 3.8.

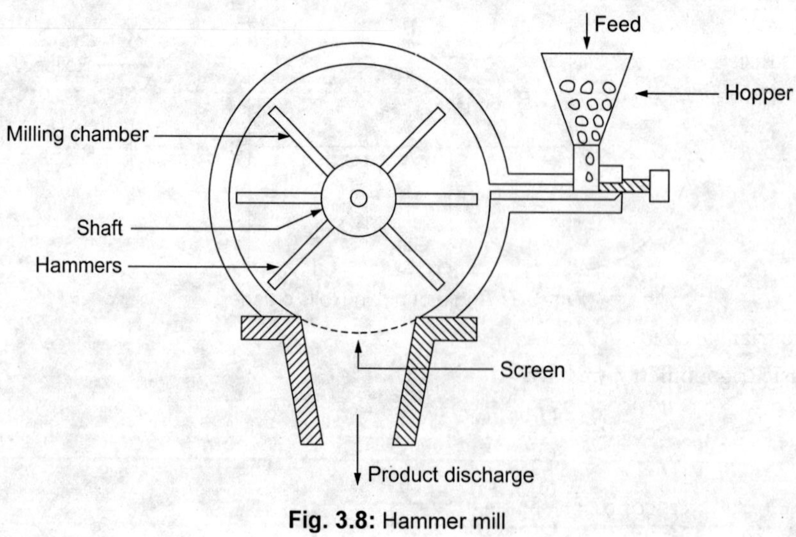

Fig. 3.8: Hammer mill

The hammers are allowed to be rotate at 10,000 RPM. The feed material is place into the hopper which flows down, while hammers are in continuous motion. These rotating hammers beat the materials to yield smaller particles. Then, these smaller particles pass through the screen. Due to the tangential exit, the size of the product is considerably smaller than the aperture of the screen. The screens are interchangeable, so that any grade of fineness can be achieved. The hammer act as a centrifugal fan, so that large amount of air is drawn through the mill. In most of the cases, this is sufficient to counteract the heat generated during milling.

It is used in pharmaceutical industries to process wet or dry granulations and disperse powder mixtures.

3.11.2.2 Rolling Compression Mill

There are plenty of rolling compression mill. But, here discussed only two types:

(*i*) Bowl mill

(*ii*) Roller mill

Bowl mill: Bowl mills are employed to pulverize the pre-crushed raw coal to the required fineness before it is admitted into the boiler furnace for combustion. The Bowl mill is driven by AC motor, coupled with flexible coupling to the worm shaft. The worm shaft drives worm gear, mounted on the main vertical shaft. A bowl mounted on the top of shaft rotates at a speed of 50 RPM. A typical Bowl mill is shown in Fig. 3.9.

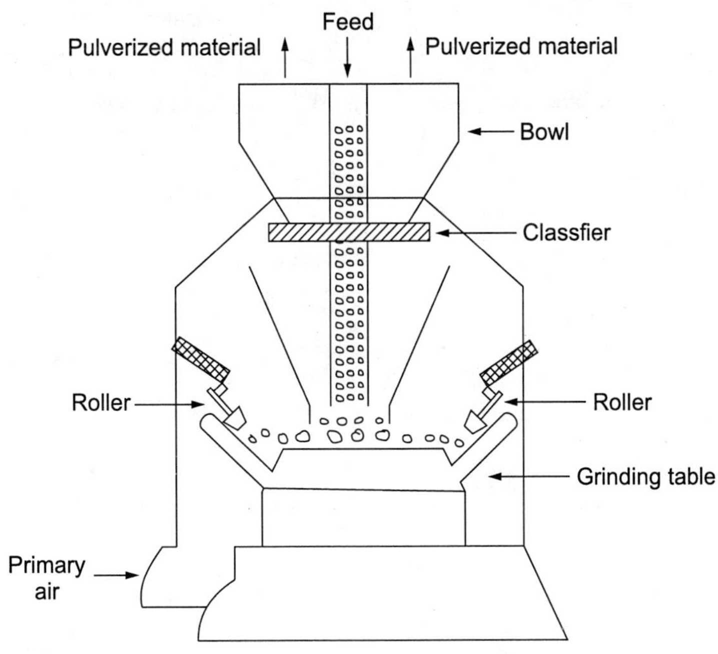

Fig. 3.9: Bowl mill

For a particular mill the speed is constant. Integral coolers are provided in the mill base with circulating water for removing heat from the oil. External lube oil systems are also used for cooling the oil in the gear box, which consists of a motor driven pump, cooler and filter. To prevent coal entering into the gear case, an air sealing system is provided. Hot primary air for drying and carrying pulverized coal entered through the insulated mill side and linear assembly. These are scraped and dumped in the pyrite hopper through the spout valve. The journal assembly consisting of grinding roll is mounted on separator body. Grinding force is applied by pressure spring adjustment. The roller assembly can be moved up and down depending upon the roll condition and coal bed. Separator body consisting of linear assembly deflects the air stream to the center of the bowl. Mill is equipped with the stationery classification with adjustable vanes for adjustment of coal are returned back to the bowl for further grinding. Mill discharge valves mounted on the separator top consists of flap type closure plates operated by pneumatic cylinder, which can be tilted out of the path of coal stream. This is used to isolate the mill from boiler during the maintenance.

Roller mill: It is a form of compression mill that uses a single, double or triple cylindrical heavy wheel mounted horizontally and rotated about their long axis either in opposing pairs or against flat plates to grind various materials. One of the roller is run by a motor and the others are rotated by friction as the material is drawn through the gap between the rollers.

The roller mill use the process of stress, which is applied by the rotating wheels and attrition in milling of solids in suspensions, pastes or ointments and some solid materials. A typical roller mill is shown in Fig. 3.10.

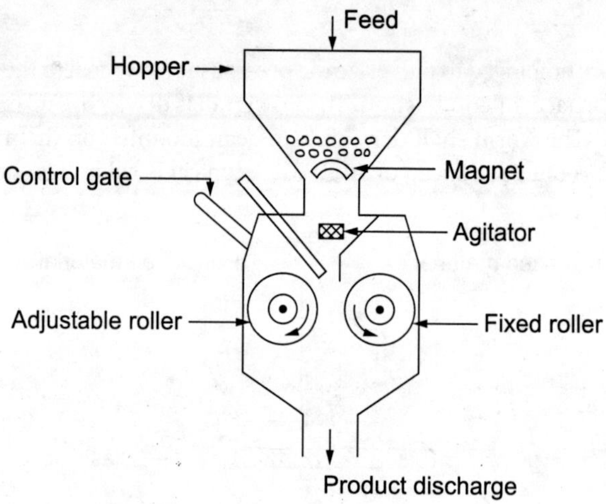

Fig. 3.10: Roller mill

The rollers rotate at different speeds and the material is crushed as it pass through the gap. To obtain the desired particle size the gap between the roller and its speed should be controlled. The design of a roller mill differs depending upon its application and they are usually named according to the work they perform.

3.11.2.3 Attrition Mill

Attrition mill work on the principle of shattering by impact. However, they also impart a shearing and cutting action.

The discs of an attrition mill are generally in a vertical position so that materials not capable of reduction can pass by gravity out of the grinding area. A typical attrition mill is shown in Fig. 3.11.

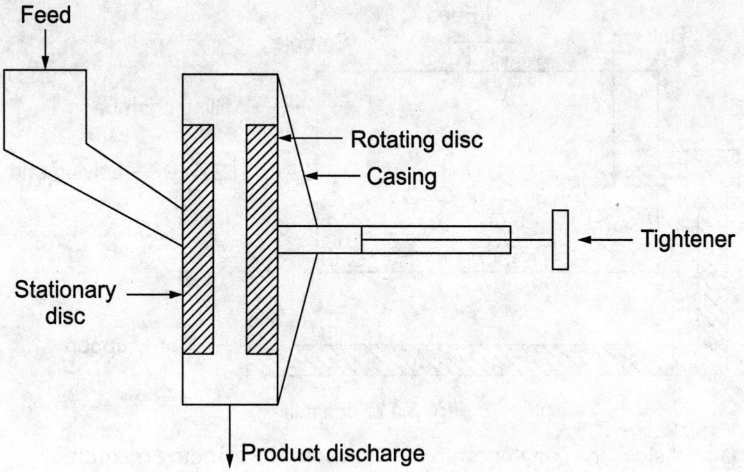

Fig. 3.11: Attrition mill

Grinding is done between the two discs equipped with replaceable wearing surfaces. One or both of these discs is rotated. If both, they rotate in opposite direction. When one disc is rotated and the other stationary, the assembly is used for shredding and deferring. Often materials which have been coarsely ground by other mills are passed through an attrition mill for blending or smoothing out an ingredient containing liquids which may have lumps. The disc of an attrition mill are generally in a vertical position so that materials not capable of reduction can pass by gravity out of the grinding area.

It is used in pharmaceutical industries for mixing and blending of fine particles.

3.11.2.4 Tumbling Mills

Tumbling mills have been widely implemented in many industries for grinding of bulk materials. They have been used for decades in the production of fines and in the final stages of ore comminution where optional levels for the enrichment particle size are obtained. There are plenty of Tumbling mills. But, only a few are discussed here.

(*i*) Ball mill

(*ii*) Rod mill

(*iii*) Tube mill

Ball mill: Ball mill work on the principle of impact and attrition i.e. size reduction take place as the ball drops from near the top of a rotating hollow cylindrical shell.

The Ball mill consists of a hollow cylinder which is mounted on a metallic frame in such a way that it can be rotated on its longitudinal axis. The length of the cylinder is slightly greater than its diameter. The cylinder is made of a metal and is usually lined with an abrasion resistant material such as manganese steel or rubber. The cylinder contains stainless steel balls that occupy about 40% of the mill volume. The weight of the ball is kept constant. The ball size depends on the size of feed and the diameter of the mill. A typical Ball mill is shown in Fig. 3.12.

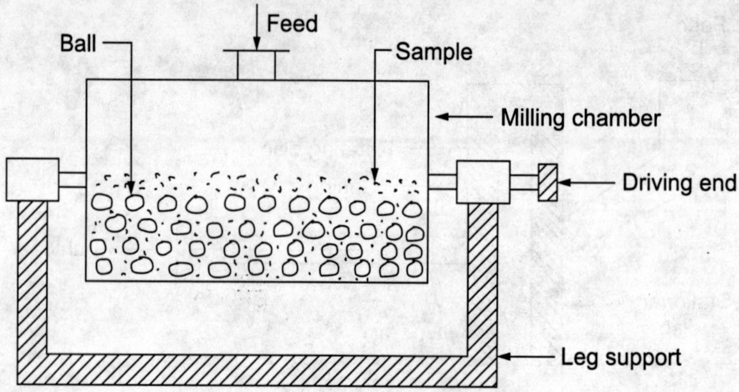

Fig. 3.12: Ball mill

The balls are made of steel, iron or stoneware. These act as the grinding medium.

The material to be ground is put into the cylindrical shell of the mill in such a quantity that it is filled about 50% of the total volume. A fixed number of balls are introduced and the cylinder is closed. The mill allowed settling to rotate on its longitudinal axis.

The speed of rotation is very important. At low speed, the balls roll over each other and attrition will be predominant mode of stress. The use of small ball is recommended so that the surface is the greatest.

This mode of attrition is used for wet grinding. It may be useful to add surface active agent to prevent agglomeration.

At higher speed, the centrifugal force just occurs as a result the mill wall picks up the balls and carried nearly to the top, where they break contact with the wall and fall to the bottom to picked up. In this manner, impact stress will also be induced and the size reduction is made effective.

At still higher speed the ball are thrown out to the wall by centrifugal force. Hence, grinding will not occur. The compression by the balls against the wall will not be sufficient for effective communication of the particles. The modes of rolling of balls in a ball mill are shown in Fig. 3.13.

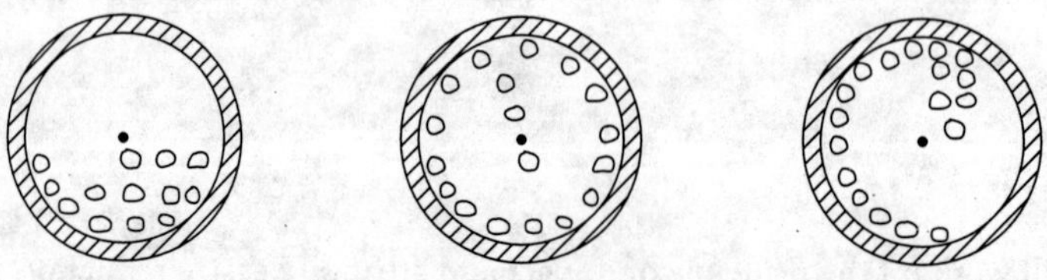

(*a*) Low speed (*b*) High speed (*c*) Correct speed

Fig. 3.13: Modes of rolling of ball in a Ball mill

The grinding with a particle size 5 mm to 50 mm or less can be obtained particularly for hard abrasive materials.

The Ball mill is used for grinding materials such as coal, pigments and feldspar for potter. Grinding can be carried out either wet or dry.

The ball mill offers several advantages. These are:

(*i*) It can produce very fine powder.

(*ii*) It is used for batch operation.

(*iii*) Toxic substances can be ground, as it is a closed system

(*iv*) Milling operation can be accomplished in an inert atmosphere.

(*v*) Ball mill can be of various shape and size.

(*vi*) Installation operation is low.

Critical speed of the Ball mill: During the operation, the mill is rotated and the balls are picked up by the wall and carried nearly to the top, where they break contact with the wall and fall to the bottom to be picked up again. Centrifugal force keeps the balls in contact with the wall and with one another during the upward movement. While in contact with the wall, the balls do some grinding by slipping and rolling over one another, but most of the grinding occur at zone of impact, where the free falling balls strike the bottom of the mill.

The balls are projected across the mill depending upon the speed of rotation. At low speed of operation, the balls simply roll over each other resulting into little grinding action. When the mill is operated at higher speed, the balls will be carried up further inside the mill and greater will be the power consumption. But at the same time, as the balls fall down from the higher distance, greater will be the impact at the bottom and larger will be the crushing.

If the mill is operated at very high speed, the balls are carried round in contact with side of the mill and the mill is said to be centrifuging. The minimum speed at which centrifuging occurs is called critical speed of the Ball mill. Under these conditions centrifugal force will be exactly balanced by the weight of the ball. Either little or no grinding takes place when a mill is centrifuging. For better performance of the mill, the operating speed must be less than the critical speed.

The speed at which the outermost ball break contact with the wall depends on the balance between the centrifugal and gravitational forces. Forces on ball in ball mill is shown in Fig. 3.14.

Consider a ball at point *B* on the periphery of the Ball mill. Let *R* be the radius of the mill and *r* be the radius of the ball. *R-r* represent the distance between the axis of the mill and the centre of the ball. Let α be the angle between *OB* and the vertical *OC*. The forces acting on the ball are:

(*i*) The force of gravity = $mg \cos \alpha$ ...(3.16)

where, m = Mass of the ball

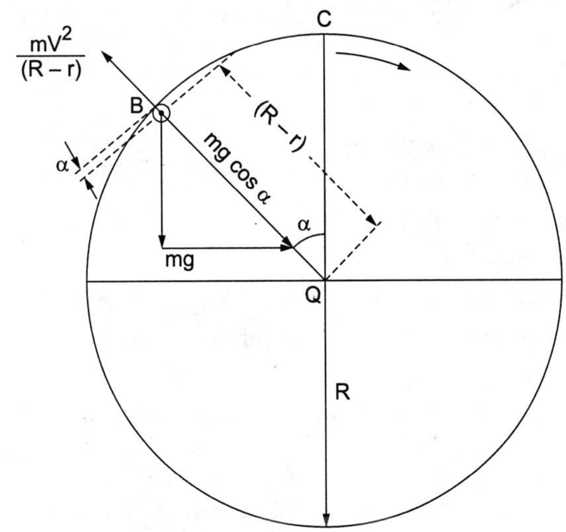

Fig. 3.14: Forces on ball in Ball mill

(*ii*) The centrifugal force $= \dfrac{mV^2}{(R-r)}$ ...(3.17)

where, V = Peripheral speed

The centripetal component of the force of gravity $mg \cos \alpha$ opposes the centrifugal force. As long as the centrifugal force exceeds the centripetal component of the force of gravity, the ball will be loose contact with the wall of the mill. As the angle α decreases, the centripetal force increases and unless the speed crosses the critical value, a stage is reached where the opposing forces are equal and the ball is ready to fall down at the bottom of the mill. The forces at this phenomena occurs can be found by comparing the equation (3.15) and equation (3.16).

$$\Rightarrow \quad mg \cos \alpha = \frac{mV^2}{(R-r)} \qquad ...(3.18)$$

$$\Rightarrow \quad \cos \alpha = \frac{V^2}{(R-r)g} \qquad ...(3.19)$$

The relation between peripheral speed and the speed of rotation is given by the expression:

$$V = 2\pi N(R-r) \qquad ...(3.20)$$

Substituting equation (3.20) in equation (3.19) we get

$$\cos \alpha = \frac{4\pi^2 N^2 (R-r)}{(R-r)g}$$

$$\Rightarrow \quad \cos \alpha = \frac{4\pi^2 N^2 (R-r)}{g} \qquad ...(3.21)$$

At critical speed $\alpha = 0$, $\cos \alpha = 1$ and N becomes the critical speed N_C, the equation (3.21) reduces to

$$1 = \frac{4\pi^2 N_c^2 (R-r)}{g} \qquad ...(3.22)$$

$$\Rightarrow \quad N_c^2 = \frac{g}{4\pi^2 (R-r)}$$

$$\Rightarrow \quad N_c = \frac{1}{2\pi}\sqrt{\frac{g}{(R-r)}} \qquad ...(3.23)$$

where, N_c = Critical speed of the Ball mill

g = Acceleration due to gravity, m/s^2

R = Radius of the Ball mill, m

r = Radius of the Ball, m

The operating speed of the Ball mill is between 50% to 75% of the critical speed.

Rod mill: Rod mill work on the principle of impact and attrition. It is similar to Ball mill except they uses long rods in the place of balls as grinding media.

The rods grind the ore by tumbling with the mill similar to the grinding balls in Ball mill. To prevent the conditions lead to rod charge tangling, the length to diameter ratio of the Rod mill is maintained as 1.4 to 1.6. The Rod mill accept feed upto about 50 mm and produces a product in the size range of 3 mm to 0.3 mm. Grinding action is by line contact between the rod extending the length of the mill. Rods tumble and spin roughly parallel alignment simulating a series of roll crushers. A typical Rod mill is shown in Fig. 3.15.

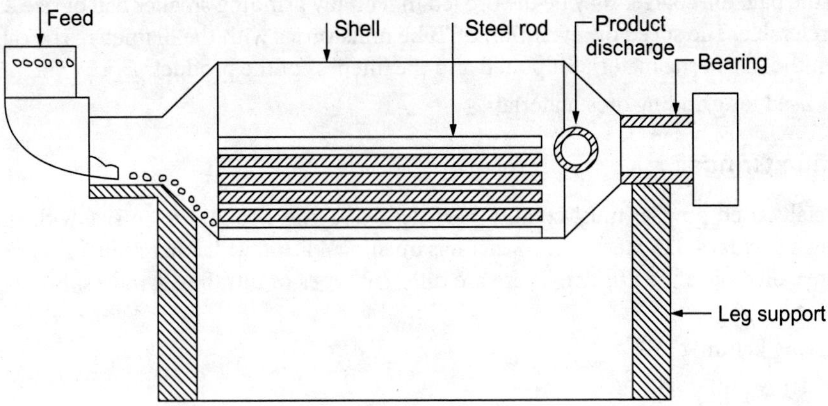

Fig. 3.15: Rod mill

It results in preferential grinding of coarse material and minimizes the production of slimes. Wet grinding rod mills are normally used in the mineral processing industries. Dry grinding is used in some areas, however it is confronted with problem and should be avoided as far as possible, except where absolutely necessary.

Rod mill operate at lower speed than the Ball mill since the rods are rolled and not cascaded. It is important that the rod stay essentially parallel to one another inside the mill. If the rod become misaligned, grinding action is lost and more importantly rod tangles occurs.

The rod mills normally carry 35 to 65% rod charges by volume and it operates between 50 to 95% of the critical speed.

Tube mill: Tube mill consists of a revolving cylinder, the interior of which is perfectly plain. They contain a huge number of hard steel balls. They differ from ball mills essentially in having the inlet for ore at one end and the outlet at the other end. There is a provision in the mactime itself for the return of uncrushed oversize material. Therefore, it is necessary for the cylinder to be of greater length than in Ball mills. The number of balls used is also much larger. According to one view the reduction of the ore is effected by grinding rather than by impact.

The cylinder are from 3 to 6 feet in diameter and from 13 to 25 feet long. The lining consists of hard wood, steel, manganese steel, quartzite or flint. Among these manganese lining seems to be the best one. But, the greatest cost of manganese steel made it less economical than hard iron. The balls are made of steel. The size of the ball varies from 75 mm to 100 mm in diameter. The mill is best operate at half filled with steel balls. A typical Tube mill is shown in Fig. 3.16.

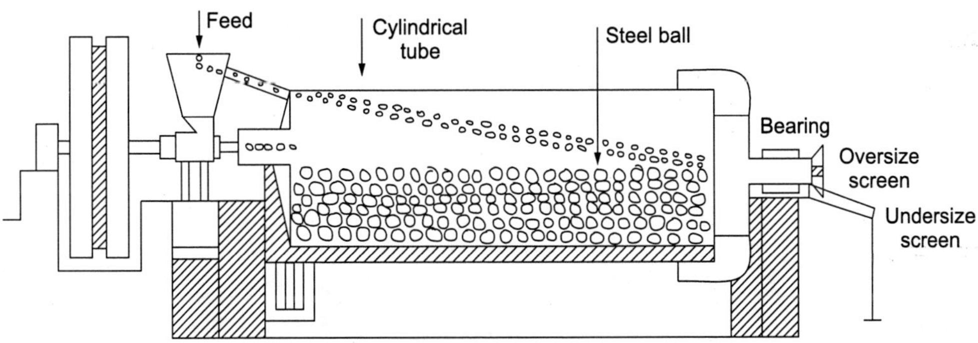

Fig. 3.16: Tube mill

The larger the ball, the coarser may be the ore fed in. For any grinding smaller ball of size 25 to 50 mm in diameter are preferable. The speed of revolution of Tube mills varies with the diameter. The capacity of Tube mill depends on the size of the material supplied and the fineness of the product.

Rod mill is used for grinding ores materials.

3.11.3 Ultrafine Grinder

Many commercially used powder must contain particles averaging 1 to 20 μm in size with substantially all particles passing a standard 325 mesh screen that has opening 44 μm. Mills that reduces the solids to such a fine particles are called ultrafine grinder. There are different types of ultrafine grinders. But, here only a few types are discussed.

(*i*) Classifying hammer mill
(*ii*) Fluid energy mill
(*iii*) Colloidal mill
(*iv*) Agitator mill
(*v*) Edge runner mill
(*vi*) End runner mill

3.11.3.1 Classifying Hammer Mill

It work on the principle of impact and attrition.

In this mill a set of swinging hammer is held in between the two rotor disk. The rotor shaft also carries two fan, which draw air through the mill and discharge into ducts leading to collectors for the product. There are short radial vanes on the rotor disk, which is used for separating oversize particles from those of acceptable size. A typical classifying Hammer mill is shown in Fig. 3.17.

In grinding chamber the solid particles are given high rotation. Therefore, the coarse particles are concentrated along the wall of the chamber due to centrifugal force acting on them. The air stream carries fine particles inward from the grinding chamber towards the shaft. The separator vane tends to throw particles outwards. The chances of particle passing between the classifier and out to the discharge depends upon the force that dominates the drag exerted by the air or the centrifugal force exerted by the vanes. The fine particles are carried through the vanes and the large particles are thrown back in the grinding chamber for further reduction.

The feed size in classifying Hammer mill is 50-5 mm and the product size can be obtained as 5-0.1 mm.

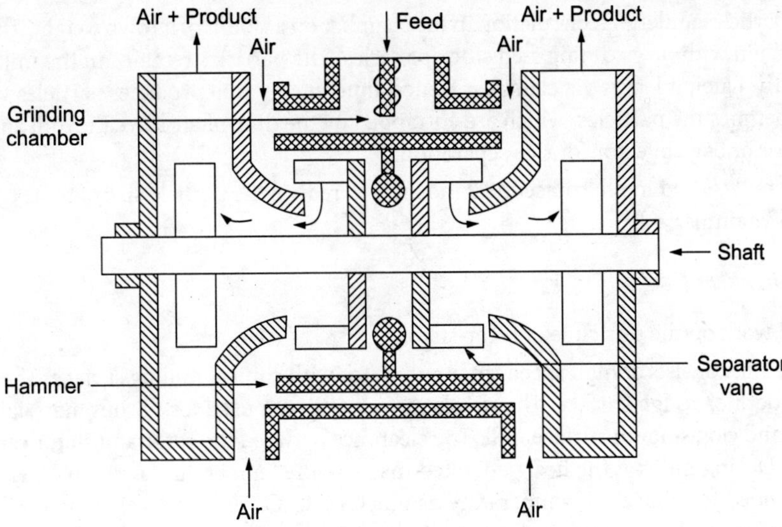

Fig. 3.17: Classifying Hammer mill

3.11.3.2 Fluid Energy Mill

It operates on the principle of impact and attrition.

Fluid energy mill consists of an elliptical pipe, which has height of about 2 m and diameter may be around 20 to 200 mm. The mill surface may be made of either stainless steel or tough ceramics. Usually, the mills constructed such that the contact surfaces are merely linings, which can be removed or replace, if excessively eroded after use. Grinding nozzles may placed tangential flow path of the particles. Normally, compressed air is used at 5-10 MN/m^2. Inert gases can be used to minimize the oxidation of susceptible compounds. Venturi feeder is provided in the path of the airflow. An outlet with a classifier is fitted to allow the escape of air. A typical Fluid energy mill is shown in Fig. 3.18.

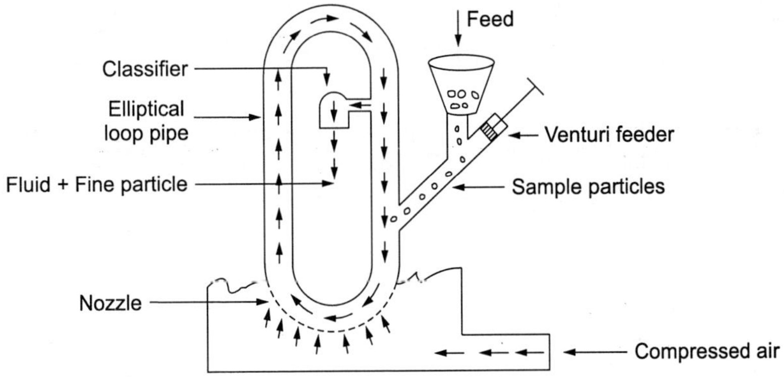

Fig. 3.18: Fluid energy mill

The sample is introduced through the inlet of venturi. The air enters through the grinding nozzles transport the sample in the elliptical path of the mill. In the turbulent stream of air, the suspended particles collide with each other and break. Thus, impact and attrition forces operate in size reduction. The resultant small particles are carried to the outlet and removed by cyclone or filters.

The coarser particles undergo recirculation in the chamber on account of its own weight. These recirculated particles collide again with new coming feed stock particles. The particles remains in the mill, until its size is reduced sufficiently. Later, it leaves via classifier. Hence, fluid energy mill produces particles with narrow size distribution. Sometimes the particles, which are entrapped by the drag of air leave the mill and carried out to a cyclone separator or bag collector for size separation.

Fluid energy mill is used in pharmaceutical industry to reduce the particle size of most of the drugs such as antibiotics and vitamins.

3.11.3.3 Colloidal Mill

The colloidal mill work on the principle of rotor-stator.

The colloidal mill consists of high speed rotor and stator with conical milling surface. The milling surface may be either smooth or rough surface. The rough surface mills are used for fibrous material, because fibers tend to interlock and clog smooth surface mill. The clearance between the rotor and stator can be adjusted to any requirement. During milling, the heat generated may rise the temperature upto 60°C. Hence, cold water circulation is provided to reduce the temperature as much as 25°C. The discharge pipe is also connected to hopper so that discharge can be recycled. A typical Colloidal mill is shown in Fig. 3.19.

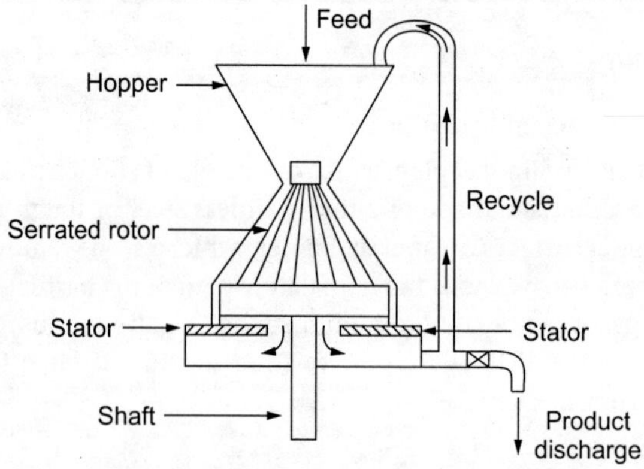

Fig. 3.19: Colloidal mill

Material such as suspensions and emulsions are placed in the hopper. The solids are mixed with liquid before being introduced into the colloid mill. The dispersions flow down and adheres to the rotor. During the movement of rotor, centrifugal forces throw a part of the dispersion on to the rotor. Depending on the clearance, the dispersion is sheared between the rotor and stator. Thus, size reduction is achieved. The milled liquid may be recycled. After, achieving the desired size, the discharge is collected from the outlet in the periphery of the mill.

Normally, the size of the milled particles may be smaller than the clearance, because of the force of high shear. In emulsification a clearance of 75 μm may produce dispersion with an average particle size of 3 μm.

It is used for the preparation of ointment.

3.11.3.4 Agitator Bead Mill

It work on the principle of compression and shear force.

An Agitator bead mill consists of a grinding chamber of a rotating shaft equipped with agitator elements, a motor drive and a media separator. The agitator elements are typically disks or pins. A typical Agitator Bead mill is shown in Fig. 3.20.

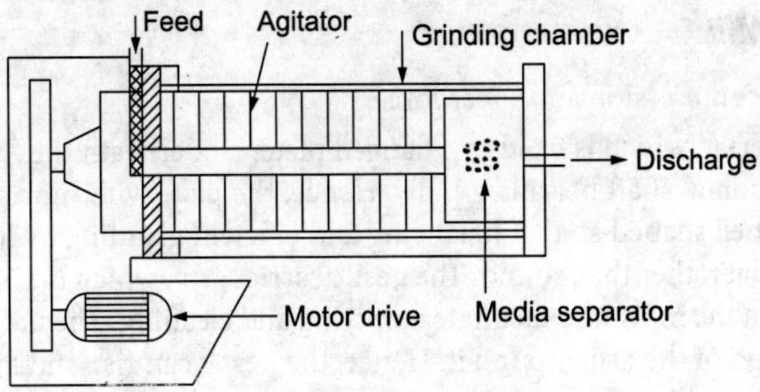

Fig. 3.20: Agitator Bead mill

The grinding chamber is filled with 9000-30 microns grinding media upto 95% of the mill volume. The grinding media can be made from the materials such as stainless steel and glass.

In operation the suspension containing the coarse material is pumped into the mill from the feed tank. The material flows into the grinding chamber downward in the space between the grinding media. As the agitator rotates, the media move around the chamber and impart compression and shear forces to the suspended particles, fracturing or dispersing them. The particles reduced to the required fineness and discharge through the media separator to a product tank.

The Agitator bead mill can be used to obtained nanoparticle dispersion.

3.11.3.5 Edge Runner Mill

It work on the principle of compression and shear force.

It consists of two heavy rollers. The weight of the roller may be several tons. The rollers move on a bed, which is made of stone. Each roller has a central shaft and revolve on its axis. Further, the rollers are mounted on a horizontal shaft and move around the bed. A typical Edge runner mill is shown in Fig. 3.21.

The material to be ground is placed on the bed with the help of a scrapper, it is kept in the path of the stone wheel. The stone rollers revolve on its axis and at the same time travel round the shallow stone bed. The outer part of the wheel has to travel a greater distance than the inner, so that size reduction achieved by shearing as well as crushing. The material is ground for a definite time period. The product is collected and passed through a sieve to get required particle size.

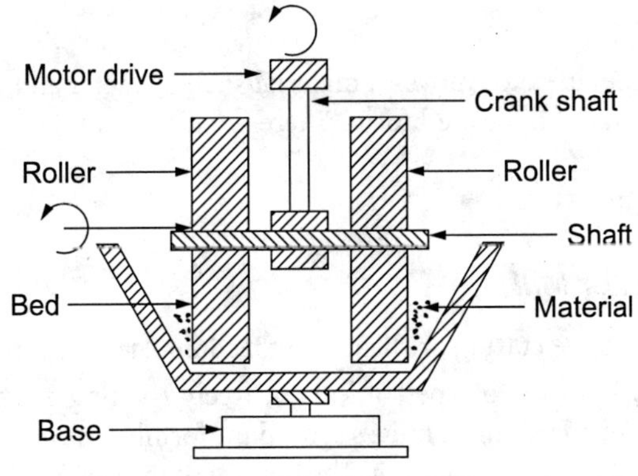

Fig. 3.21: Edge runner mill

The Edge runner mill is used for grinding tough materials to fine powder.

3.11.3.6 End Runner Mill

It work on the principle of compression and shear force.

It consists of a steel mortar, which is fixed to a flanged plate. Underneath the flanged plate, a beveled cog fitting is attached to a horizontal shaft bearing a pully. Hence, the plate with mortar can be rotated at a high speed. The pestle is dumb-bell shaped so that balancing and efficient grinding by its weight can be achieved. The bottom of the pestle is flat rather than round. The pestle carries arm, which is hinged. By this arrangement, the pestle can be raised from the mortar to facilitate emptying and cleaning. The narrow central portion of the pestle is longer than the band of the arm around it. Hence, the pestle can rise and fall over the material in the mortar. A typical End runner mill is shown in Fig. 3.22.

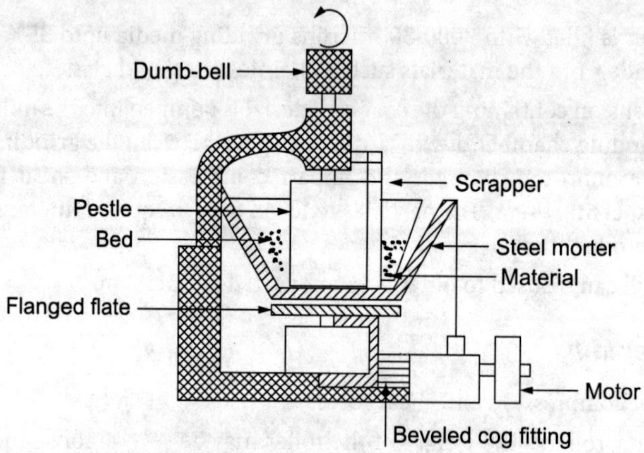

Fig. 3.22: End runner mill

The material to be ground is placed in the mortar. The scrapper puts the material in the path of the pestle. The mortar revolves at a high speed. The pestle is placed in the motor. The revolving mortar causes the pestle to revolve. During the process, size reduction is achieved by crushing as well as shearing. The crushed material is collected and passed through a sieve to get the desired particle size.

The End runner mill is used for fine grinding.

3.11.4 Cutting Machines

Cutting is a technique where the operator moves a material such as metal and the tool in a relation to each other in order to obtain the material in desired form. There are different types of cutting machine. But, here discussed only on:

(*i*) Rotary knife cutter mill

3.11.4.1 Rotary Knife Cutter Mill

It work on the principle of successive cutting or shearing of feed material with the help of sharp knives.

The milling chamber consists of two types of knives, freely rotating knives and stationary knives. A horizontally mounted rotor disk consists of 12 knives spaced uniformly. The casing also has several stationary knives. The hopper is placed at the top of the casing. The bottom of the casing holds a screen that controls the size of the material. Discharge opening is provided at the bottom of the mill. A typical Rotary knife cutter mill is shown in Fig. 2.23.

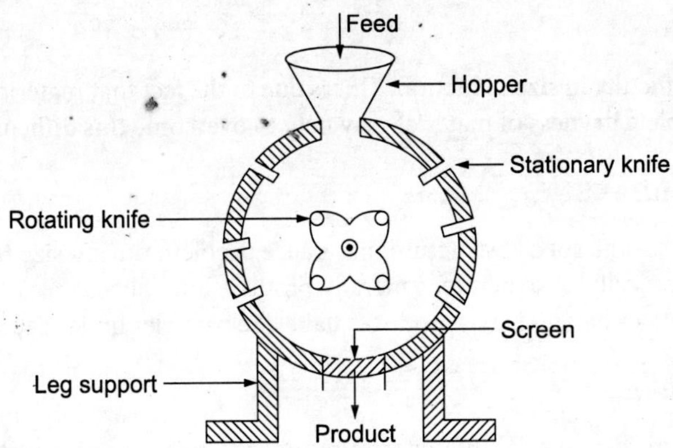

Fig. 3.23: Rotary knife cutter mill

The rotor disk is allowed to rotate at speed from 200-100 RPM. The feed material is loaded into the hopper, which flows down by the gravity. During the rotation of the disc, the materials comes very close between the stationary and rotating knives, thereby the material is cut into small pieces. Smaller particles pass through the screen. The knives lift the coarser particles up while rotating and promote further size reduction. The product is collected into a receiver.

The Rotary knife cutter mills are used for the size reduction of tough and fibrous materials. Medicinal plants are normally converted into small pieces using cutter mill. Soft materials such as roots, peels and wood are cut into small pieces before extraction. It is also used in the manufacture of rubber, plastics, recycling of paper waste materials.

3.12 FACTOR AFFECTING SIZE REDUCTION

The following factors affect the process of size reduction.

(*i*) Hardness
(*ii*) Toughness
(*iii*) Stickness
(*iv*) Material structure
(*v*) Moisture content
(*vi*) Softening temperature
(*vii*) Purity required
(*viii*) Ratio of feed size to product size
(*viii*) Bulk density

3.12.1 Hardness

The hardness of the material affects the process of size reduction. It is easier to break soft material to a small size rather than hard material.

3.12.2 Toughness

The crude material of fibrous nature or those material having higher moisture content are generally tough in nature. A soft but tough material may cause more problem in size reduction than a hard but brittle substances.

3.12.3 Stickness

Stickness causes a lot of difficulty in size reduction. This is due to the fact that material adheres to the crushing surfaces of the mill. Complete dryness of material may help to overcome this difficulty.

3.12.4 Material Structure

The materials which shows some special structure may cause problem during size reduction e.g., substances from vegetable sources have cellular structure, generally produce long fibrous particles on its size reduction. Similarly, a mineral substances having flaws, produces flake like particles on its size reduction.

3.12.5 Moisture Content

The presence of moisture in the material influences a number of its properties such as hardness, toughness and stickness which in turn affects the particle size reduction. The material should be either dry or wet. It should not be damp. The material having 5% moisture in case of dry grinding and 50% moisture in wet grinding does not create any problem.

3.12.6 Softening Temperature

The substance containing oil or fat, become softened during the size reduction processes if heat is generated. This can be avoided by cooling the mill.

3.12.7 Purity Required

Many of the mills used for size reduction often cause the grinding surfaces to wear off and thus impurities comes in the product sample. If a high degree of purity is required, such mills must be avoided. Moreover, the mills should be thoroughly cleansed in order to maintain purity.

3.12.8 Ratio of Feed Size to Product Size

To get a fine particles in a mill, it is required that a fairly small feed size should used. Hence, it is necessary to carry out the size reduction process in several stages, using different mills, e.g., primary crushing followed by coarse particles and then fine grinding.

3.12.9 Bulk Density

The output of the size reduction of material in a machine depends upon the bulk density of the substance.

3.13 SIZE ENLARGEMENT

To improve the handling properties, reduce the pressure drop in fluidized beds, increase the rate of settling or drying or prevent dusting small particles often binded together to form a lumps of the desired size.

This may be accomplished by forming small pellets or larger briquettes by compacting, by fusion through a die under high pressure, by solidification of liquid drops in a spray tower or prilling tower. By flocculation and coagulation of particles in a dilute slurry or by the bonding in nodulization and sintering processes.

It is difficult to predict the behavior of a given material without actual tests. A binder is often added when making pellets, briquettes or extrudates so that the particles are stick together.

Example 3.1. A certain crusher accepts a feed material having volume surface mean diameter 20 mm and produces a product of volume surface mean diameter of 5 mm. Calculate the power required to crush 15 t/hr of feed material. Assume, $K_r = 3.45 \times 10^{-3}$.

Solution: From the Rittinger's law, we have

$$\frac{P}{\dot{m}} = K_r\left[\frac{1}{D_{VSP}} - \frac{1}{D_{VSF}}\right]$$

Data given:

$$\dot{m} = 15 \text{ t/hr}$$
$$K_r = 3.45 \times 10^{-3}$$
$$D_{VSP} = 5 \times 10^{-3} \text{ m}$$
$$D_{VSP} = 20 \times 10^{-3} \text{ m}$$

$$\Rightarrow \quad \frac{P}{15} = 3.45\times10^{-3}\left[\frac{1}{5\times10^{-3}} - \frac{1}{20\times10^{-3}}\right]$$

$$\Rightarrow \quad P = (3.45\times10^{-3})\times15\times\left[\frac{1}{5\times10^{-3}} - \frac{1}{20\times10^{-3}}\right]$$

$$\Rightarrow \quad P = 7.76 \text{ kW}$$ **Ans.**

Example 3.2. Particle of the average feed size of 50×10^{-4} m are crushed to an average product size of 10×10^{-4} m at the rate of 20 t/hr. At this rate, the crusher consumes 40 kW of power of which 5 kW are required for running the mill empty. Calculate the power consumption if 10 t/hr of this product is further crushed to 5×10^{-4} m size in the same mill. Assume Rittinger's law.

Solution: From the Rittinger's law, we have

$$\frac{P}{\dot{m}} = K_r\left[\frac{1}{\bar{D}_{VSP}} - \frac{1}{\bar{D}_{VSF}}\right]$$

Data given:

$$\bar{D}_{VSP} = 10 \times 10^{-4} \text{ m}$$
$$\bar{D}_{VSF} = 50 \times 10^{-4} \text{ m}$$
$$\dot{m} = 20 \text{ t/hr}$$
$$P = 40 \text{ kW, for crushing the sample}$$
$$P = 5 \text{ kW, for running empty mill}$$
$$\text{Actual power consumption} = 40 - 5 = 35 \text{ kW}$$

$$\Rightarrow \quad \frac{35}{20} = K_r\left[\frac{1}{10\times10^{-4}} - \frac{1}{50\times10^{-4}}\right]$$

$$\Rightarrow \quad 1.75 = K_r\left[\frac{1}{0.001} - \frac{1}{0.005}\right]$$

$$\Rightarrow \quad 1.75 = K_r[1000 - 200]$$

$$\Rightarrow \quad K_r = 2.187 \times 10^{-3} \text{ kWh.m/t}$$

This value of K_R is constant for the crusher.

Data Given:

$$\bar{D}_{VSF} = 10 \times 10^{-4} \text{ m}$$
$$\bar{D}_{VSP} = 5 \times 10^{-4} \text{ m}$$
$$\dot{m} = 10 \text{ t/hr}$$

$$\Rightarrow \quad \frac{P}{10} = 2.187\times10^{-3}\left[\frac{1}{5\times10^{-4}} - \frac{1}{10\times10^{-4}}\right]$$

$$\Rightarrow \quad \frac{P}{10} = 2.187 \times 10^{-3}[2000 - 1000]$$

$$\Rightarrow \quad P = 21.87 \text{ kW} \quad \textbf{Ans.}$$

Example 3.3. A sample of material is crushed in a Blake Jaw crusher such that the average size of the particles is reduced from 40 mm to 10 mm with energy consumption of 13 kW/(kg/s). Determine the consumption of energy to crush the same material of 75 mm average size to an average size of 25 mm using Rittinger's law and Kick's law.

Solution: Rittinger's law:

Data Given:

$$\bar{D}_{VSF} = 40 \text{ mm}$$

$$\bar{D}_{VSP} = 10 \text{ mm}$$

$$\frac{P}{\dot{m}} = 13 \text{ kW/(kg/s)}$$

$$\frac{P}{\dot{m}} = 3.61 \text{ kWh/t}$$

$$\frac{P}{\dot{m}} = \left[\frac{1}{\bar{D}_{VSP}} - \frac{1}{\bar{D}_{VSF}}\right]$$

$$\Rightarrow \quad 3.61 = \left[\frac{1}{10} - \frac{1}{40}\right]$$

$$\Rightarrow \quad 3.61 = K_R[0.1 - 0.025]$$

$$\Rightarrow \quad K_R = 48.13 \text{ kWh.mm/t}$$

When the same machine is used for crushing of particle size from 75 mm to 25 mm.

$$\frac{P}{\dot{m}} = 48.13\left[\frac{1}{25} - \frac{1}{75}\right]$$

$$\Rightarrow \quad \frac{P}{\dot{m}} = 48.13[0.04 - 0.013]$$

$$\Rightarrow \quad \frac{P}{\dot{m}} = 1.2995 \text{ kWh/t} \quad \textbf{Ans.}$$

Kick's Law:

$$\frac{P}{\dot{m}} = K_k \ln\left[\frac{\bar{D}_{VSF}}{\bar{D}_{VSP}}\right]$$

$$\Rightarrow \quad 3.61 = K_k \ln\left[\frac{50}{10}\right]$$

$$\Rightarrow \quad 3.61 = K_K \ln [5]$$

$$\Rightarrow \quad K_K = 2.243 \quad \textbf{Ans.}$$

When the same machine is used for crushing of particles size from 75 mm to 25 mm.

$$\frac{P}{\dot{m}} = 2.243 \ln\left(\frac{75}{25}\right)$$

$$\Rightarrow \quad \frac{P}{\dot{m}} = 2.243 \times \ln(3)$$

$$\Rightarrow \quad \frac{P}{\dot{m}} = 2.4642 \text{ kWh/t} \quad \textbf{Ans.}$$

Example 3.4. 250 kW of power is required to crush 150 t/hr of a material. If 80% of the feed passes through a 50 mm screen and 80% of the product passes through a 3 mm screen. Calculate the work index of the material.

Solution: From the Bond's law, we have

$$\frac{P}{\dot{m}} = 0.3162W_i\left[\frac{1}{\sqrt{D_{PP}}} - \frac{1}{\sqrt{D_{PF}}}\right]$$

Data Given:

$$\dot{m} = 150 \text{ t/hr}$$
$$P = 250 \text{ kW}$$
$$D_{PP} = 3 \text{ mm}$$
$$D_{PF} = 50 \text{ mm}$$

$$\Rightarrow \quad \frac{250}{150} = 0.3162 \times W_i \times \left[\frac{1}{\sqrt{3}} - \frac{1}{\sqrt{50}}\right]$$

$$\Rightarrow \quad 1.667 = 0.3162 \times W_i \times \left[\frac{1}{1.732} - \frac{1}{7.071}\right]$$

$$\Rightarrow \quad 1.667 = 0.3162 \times W_i \times [0.577 - 0.141]$$

$$\Rightarrow \quad W_i = 12.09 \quad \textbf{Ans.}$$

Example 3.5. What will be the power required to crush 150 t/hr of limestone if 80% of the feed passes 50 mm screen and 80% of the product passes a 3.125 mm screen? Assume work index of limestone 12.75.

Solution: From the Bond's law, we have

$$\frac{P}{\dot{m}} = 0.3162W_i\left[\frac{1}{\sqrt{D_{PP}}} - \frac{1}{\sqrt{D_{PF}}}\right]$$

Data Given:

$$\dot{m} = 150 \text{ t/hr}$$
$$W_i = 12.75$$
$$D_{PP} = 3.125 \text{ mm}$$
$$D_{PF} = 50 \text{ mm}$$

$$\frac{P}{150} = 0.3162 \times 12.75 \times \left[\frac{1}{\sqrt{3.125}} - \frac{1}{\sqrt{50}}\right]$$

$$\Rightarrow \quad \frac{P}{150} = 0.3162 \times 12.75 \times [0.566 - 0.141]$$

$$\Rightarrow \quad P = 0.3162 \times 12.75 \times 150 \times [0.566 - 0.141]$$

$$\Rightarrow \quad P = 257 \text{ kW} \quad \textbf{Ans.}$$

Example 3.6. In a crushing operation 80% of feed passes through 50 mm screen and 80% of product passes through 5 mm screen. It was later decided to increase the product size to 80% passing through 8 mm screen. What is the percent reduction in power consumption?

Solution: From the Bond's law, we have

$$\frac{P}{\dot{m}} = 0.3162W_i\left[\frac{1}{\sqrt{D_{PP}}} - \frac{1}{\sqrt{D_{PF}}}\right]$$

Data given, Case-I:

$$D_{PP} = 5 \text{ mm}$$
$$D_{PF} = 50 \text{ mm}$$

$$\Rightarrow \quad \frac{P}{\dot{m}} = 0.3162W_i\left[\frac{1}{\sqrt{5}} - \frac{1}{\sqrt{50}}\right]$$

$$\Rightarrow \quad \frac{P}{\dot{m}} = 0.3162W_i\left[\frac{1}{2.236} - \frac{1}{7.071}\right]$$

$$\Rightarrow \quad \frac{P}{\dot{m}} = 0.3162W_i \times (0.4472 - 0.1414)$$

$$\Rightarrow \quad \frac{P}{\dot{m}} = 0.0967W_i \text{ kWh/t}$$

Data given Case-II: $D_{PP} = 8$ mm

$D_{PF} = 50$ mm

$$\Rightarrow \quad \frac{P}{\dot{m}} = 0.3162W_i\left[\frac{1}{\sqrt{3}} - \frac{1}{\sqrt{50}}\right]$$

$$\Rightarrow \quad \frac{P}{\dot{m}} = 0.3162W_i\left[\frac{1}{2.828} - \frac{1}{7.071}\right]$$

$$\Rightarrow \quad \frac{P}{\dot{m}} = 0.3162W_i(0.3536 - 0.1414)$$

$$\Rightarrow \quad \frac{P}{\dot{m}} = 0.0671W_i \text{ kWh/t}$$

Percentage reduction in power consumption

$$= \frac{\text{(Power consumed in case I)} - \text{(Power consumed in case II)}}{\text{Power consumed in case II}} \times 100$$

$$= \frac{0.0967W_i - 0.0671W_i}{0.0671W_i} \times 100$$

$$= \frac{0.0967 - 0.0671}{0.0671} \times 100$$

$= 44.11\%$ **Ans.**

Example 3.7. A certain crusher accept a feed of rock having a volume surface mean diameter 20 mm and discharge a product of volume surface mean diameter 5 mm. The power required to crush 3 kg/s of the material is 7 kW. What will be the power consumption if the capacity is reduced to 2.5 kg/s and the volume surface mean diameter of the product is changed to 4 mm? Assume the mechanical efficiency remains constant and Rittinger's law hold good.

Solution: From the Rittinger's law, we have

$$\frac{P}{\dot{m}} = K_R\left[\frac{1}{\bar{D}_{VSP}} - \frac{1}{\bar{D}_{VSF}}\right]$$

$$\Rightarrow \quad \frac{7}{10.8} = K_R\left[\frac{1}{5\times10^{-3}} - \frac{1}{20\times10^{-3}}\right]$$

$$\Rightarrow \quad \frac{7}{10.8} = K_R[200 - 50]$$

$$\Rightarrow \quad K_R = 4.32 \times 10^{-3}$$

The value of K_R is constant for the crusher.

$$\frac{P}{9} = 4.32\times10^{-3}\left[\frac{1}{4\times10^{-3}} - \frac{1}{20\times10^{-3}}\right]$$

$$\Rightarrow \quad \frac{P}{9} = 4.32 \times 10^{-3}[250 - 50]$$

$$\Rightarrow \quad P = 7.776 \text{ kWh/t} \quad \textbf{Ans.}$$

Example 3.8. Find the critical speed of the Ball mill by using the following data:

Diameter of the Ball mill, $D = 450$ mm

Diameter of the ball, $d = 25$ mm

Solution: Data Given:

Radius of the Ball mill, $R = 225 \times 10^{-3}$ m

Radius of the ball, $r = 12.5 \times 10^{-3}$ m

From the critical speed of the Ball mill, we have

$$N_c = \frac{1}{2\pi}\sqrt{\frac{g}{(R-r)}}$$

$$\Rightarrow \quad N_c = \frac{1}{2\pi}\sqrt{\frac{9.81}{(0.225-0.0125)}}$$

$$\Rightarrow \quad N_c = 64.88$$

$$\Rightarrow \quad N_c = 65 \text{ RPM} \quad \textbf{Ans.}$$

Example 3.9. Calculate the operating speed of the Ball mill from the following data:

(*i*) Diameter of the Ball mill = 600 mm

(*ii*) Diameter of the ball = 50 mm

(*iii*) Operating speed is 50% of the critical speed of the mill.

Solution: From the critical speed of the Ball mill, we have

$$N_c = \frac{1}{2\pi}\sqrt{\frac{g}{(R-r)}}$$

Data Given:

$$R = 0.30 \text{ m}$$

$$r = 0.025 \text{ m}$$

$$\Rightarrow \quad N_c = \frac{1}{2\pi}\sqrt{\frac{9.81}{(0.30-0.025)}}$$

$$\Rightarrow \quad N_c = 57 \text{ RPM}$$

Operating speed of the Ball mill = $57 \times 0.5 = 28.5 \simeq 29$ RPM

Example 3.10. What rotational speed would you recommend for a Ball mill 1000 mm in diameter charged with 75 mm diameter ball?

Solution: From the critical speed of the Ball mill, we have

$$N_c = \frac{1}{2\pi}\sqrt{\frac{g}{(R-r)}}$$

Data Given:

$$R = 0.5 \text{ m}$$

$$r = 0.0375 \text{ m}$$

$$g = 9.81 \text{ m/s}^2$$

$$\Rightarrow \quad N_c = \frac{1}{2\pi}\sqrt{\frac{9.81}{(0.5-0.0375)}}$$

$$\Rightarrow \quad N_c = 44 \text{ RPM}$$

Operating speed of the Ball mill is 50 to 75% of the critical speed.

Operating speed = 50 to 75% of 44 RPM = 22 to 39 RPM **Ans.**

Example 3.11. A pair of rolls is to take a feed equivalent sphere 40 mm in diameter and crush them to sphere having diameter 10 mm. If the co-efficient of friction is 0.29, what should be the diameter of the rolls?

Solution: From the angle of nip of roll crusher, we have

$$\cos\alpha = \frac{r+d}{r+R}$$

Data given:

$$\tan^{-1}(\alpha) = 0.29$$

$$\alpha = \tan(0.29)$$

$$\Rightarrow \quad \alpha = 16^\circ\, 17'$$

$$\Rightarrow \quad \alpha = 16^\circ$$

where, r = Radius of the roll

R = Radius of the feed particle, 20 mm

d = Radius of largest particle in the product

$$\Rightarrow \quad \cos(16) = \frac{r+5}{r+20}$$

$$\Rightarrow \quad 0.961 \times (r+20) = r+5$$

$$\Rightarrow \quad 0.961r + 19.22 = r+5$$

$$\Rightarrow \quad 0.039r = 14.22$$

$$\Rightarrow \quad r = 364.62$$

Diameter of the roll = $2r = 2 \times 364.62 = 729.23 \simeq 730$ mm **Ans.**

Example 3.12. What should be the diameter of a set of rolls to take feed size equivalent to 40 mm spheres and crush to 12 mm? The co-efficient of friction is 0.35.

Solution: From the angle of nip of roll crusher, we have

$$\cos\alpha = \frac{r+d}{r+R}$$

Data given:

$$\mu = \tan\alpha$$

$$0.35 = \tan\alpha$$

$$\alpha = \tan^{-1}(0.35)$$

$$\alpha = 19.29^\circ$$

$$\cos\alpha = 0.944$$

r = Radius of the roll, mm

R = Radius of the feed particle, 20 mm

$$d = \text{Gap between the rolls, 5 mm}$$

$$\Rightarrow \quad 0.944 = \frac{r+6}{r+20}$$

$$\Rightarrow \quad 0.944 \times (r + 20) = r + 6$$

$$\Rightarrow \quad 0.944r + 18.88 = r + 6$$

$$\Rightarrow \quad 0.056r = 12.88$$

$$\Rightarrow \quad r = 230 \text{ mm}$$

Diameter of the roll crusher = $2r = 2 \times 230 = 460$ mm **Ans.**

Example 3.13. A certain set of crushing rolls has rolls of 1200 mm diameter by 370 mm width of face. They are set so that the crushing surfaces are 12 mm apart at the narrowest point. The angle of nip is 30°. What is the maximum permissible size of the feed?

Solution: From the angle of nip of roll crushers, we have

$$\cos \alpha = \frac{r+d}{r+R}$$

Data given:

r = Radius of the roll, 600 mm

R = Radius of the feed particle?

d = Gap between the rolls/2 = 6 mm

α = Angle of nip/2

$$\alpha = \frac{30\Upsilon}{2}$$

$$\alpha = 15°$$

$$\Rightarrow \quad \cos (15) = \frac{600+6}{600+R}$$

$$\Rightarrow \quad 0.966 = \frac{600+6}{600+R}$$

$$\Rightarrow \quad 0.966 \times (600 + R) = 606$$

$$\Rightarrow \quad 600 + R = 627.33$$

$$\Rightarrow \quad R = 27.33 \text{ mm}$$

Size of the feed particles = $2 \times R = 2 \times 27.33 = 54.65 = 55$ mm **Ans.**

REVIEW QUESTIONS

3.1 Define size reduction.
3.2 Briefly explain the mechanism of size reduction.
3.3 Explain the modes of force applied in size reduction.
3.4 Explain the characteristics of comminuted product.
3.5 What is the significance of calculation of energy and power requirement for size reduction.
3.6 What are the different laws of size reduction?
3.7 Define Rittinger's law of size reduction.
3.8 Define Kick's law of size reduction.
3.9 Explain the Bond's law of size reduction.
3.10 What is work index? Explain.
3.11 Define crushing efficiency.
3.12 What is free crushing? Explain.
3.13 Define choke crushing.
3.14 Differentiate between open circuit and closed circuit grinding.
3.15 Differentiate between wet and dry grinding.
3.16 With a neat diagram explain open circuit and closed circuit grinding.
3.17 Classify the size reduction equipment.
3.18 With a neat diagram explain the construction and working of Jaw crusher.
3.19 With a neat diagram explain the construction and working of crushing roll.
3.20 With a neat diagram explain the construction and working of crushing roll.
3.21 Define angle of nip of Roll crusher.
3.22 With a neat diagram explain the construction and working of Hammer mill.
3.23 With a neat diagram explain the construction and working of Bowl mill.
3.24 With a neat diagram explain the construction and working of Attrition mill.
3.25 With a neat diagram explain the construction and working of Roller mill.
3.26 With a neat diagram explain the construction and working of Ball mill.
3.27 With a neat diagram explain the construction and working of Rod mill.
3.28 With a neat diagram explain the construction and working of Tube mill.
3.29 With a neat diagram explain the construction and working of Hammer mill.
3.30 With a neat diagram explain the construction and working of Fluid energy mill.
3.31 With a neat diagram explain the construction and working of colloidal mill.
3.32 With a neat diagram explain the construction and working of Agitator Bead mill.
3.33 With a neat diagram explain the construction and working of Edge runner mill.
3.34 With a neat diagram explain the construction and working of End runner mill.
3.35 With a neat diagram explain the construction and working of Rotary knife cutter mill.

EXERCISES

3.1 A certain crusher accepts a feed material having volume surface mean diameter 20 mm and produces a product of volume surface mean diameter of 4 mm. Calculate the power required to crush 12 t/hr of feed material. Assume, $K_R = 3.5 \times 10^{-3}$.

3.2 270 kW of power is required to crush 150 t/hr of a material. If 80% of the feed passes through a 50 mm screen and 80% of the product passes through 3 mm screen. Calculate the work index of the material.

3.3 A material is crushed in a Blake Jaw crusher and the average size of the particles is reduced from 60 mm to 20 mm with the energy consumption of 12 kW/(kg/s). What will be energy needed to crush the same material of average size from 100 mm to 25 mm, if Kick's Law is applicable?

3.4 A certain crusher takes rock whose average particle diameter is 0.025 m and crushes it to a product whose average particle diameter is 0.015 m at the rate of 2000 kg/hr. At this rate the mill takes 8.5 HP to run it empty.

(*i*) What would be the power consumption for same capacity if the average particle diameter in the product is 0.008 m?

(*ii*) How much power would be required under Kick's law condition?

3.5 A mill crushes 50 t/hr of limestone from 4 cm to 1 cm and consumes 50 kW power. Calculate the power required to reduce the size of the product from 1 cm to 5 mm for the same capacity.

3.6 A crusher accepts a feed material having a volume surface mean diameter of 18 mm and gives a product of volume surface mean diameter of 4 mm. The power required to crush 12 t/hr is 8.5 kW. What will be the power consumption if the capacity is reduced to 10 t/hr?

3.7 In a crushing operation 80% of feed passes through 48 mm screen and 80% of the product passes through 4 mm screen. It was later decided to increase the product size to 80% passing through 8 mm screen. What will be the percentage reduction in power consumption?

3.8 Calculate the operating speed of the Ball mill from the data a given below:

Diameter of the Ball mill = 800 mm

Diameter of the Ball = 60 mm

(*i*) If operating speed is 50% less than the critical speed.

(*ii*) If critical speed is 40% more than the operating speed.

3.9 What is the power required to crush 110 t/hr of limestone if 80% of the feed passes in a 50 mm screen and 80% of the product passes in a 3.175 mm screen? Work index for limestone is 12.75.

3.10 Derive the expression for critical speed of a Ball mill.

3.11 Derive the expression for angle of nip of Roll crusher.

3.12 The power required to crush 100 t/hr of material is 179.8 kW, if 80% of the feed passes through a 51 mm screen and 80% passes of product passes through a 3.2 mm screen. What is the work index of the material? What will be the power required for the same feed at 100 t/hr to be crushed a product surface that 80% is to pass through a 1.6 mm screen?

3.13 A pair of rolls is to take a feed equivalent sphere 50 mm in diameter and crush them to sphere having diameter 10 mm. If the co-efficient of friction is 0.30, what should be the diameter of the rolls?

3.14 A ball mill of 120 cm diameter is operated at 50 RPM. It is found that the mill is not working satisfactory. Explain the reasons for this and suggest a suitable operating condition. The diameter of the balls used are 70 mm. **[VTU Exam Sept. 2000]**

❑❑❑

4

CHAPTER

Flow of Fluid Past Immersed Bodies

4.1 INTRODUCTION

The laws of fluid flow and factors that control changes of pressure and velocity of fluids flowing past solid boundaries and is especially concerned with flow through closed conduits. However, the effect of the fluid on the solid is of great interest. The fluid may be either at rest or flowing past it. The situation where the solid is immersed in and surrounded by, it is immaterial which phase solid or fluid is assumed to be at rest. It is the relative velocity between the two is important. An exception to this is to met in some situation when the fluid stream has been previously influenced by solid wall and is in turbulent flow. The scale and intensity of turbulence then may be important parameter in the process.

For example, in wind tunnel the solid is at rest and the stream of air is in motion, the turbulence may give difference of forces on the solid than if the solid were moving at the same relative velocity through a quiescent and turbulence free mass of air.

4.2 DRAG

The force in the direction of flow exerted by the fluid on the solid is called drag. By Newton's third law of motion, an equal and opposite net forces is exerted by the body on the fluid. When the wall of the body is parallel with the direction of flow as in the case of a thin plate.

The only drag force is the wall shear τ_w. More generally however, the wall of an immersed body makes an angle with the direction of flow. Then, the wall of shear in the direction of flow contributes to drag. An extreme example is the drag of the flat plate perpendicular to the flow is shown in Fig. 4.1.

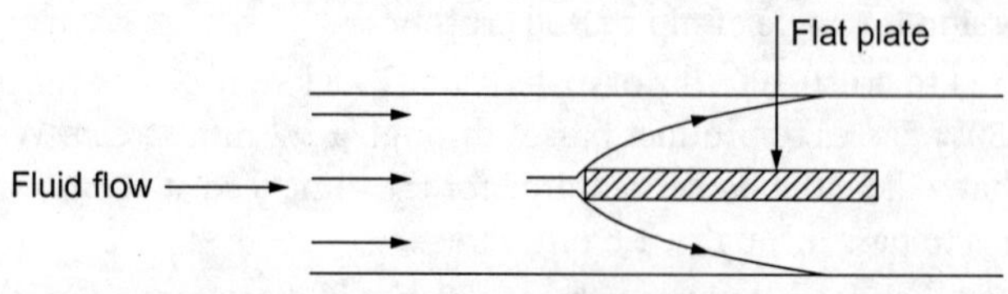

Fig. 4.1: Wall shear in the direction of flow

If the fluid pressure, which acts in the direction normal to the wall possesses a component in the direction of flow and this component also contribute to the drag. Total drag on an element of area is the sum of the two components as shown in Fig. 4.2.

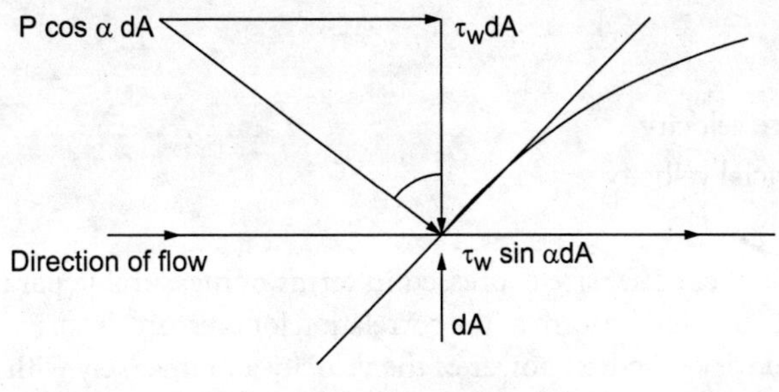

Fig. 4.2: Drag on an element

The pressure and shear forces acting on an element of area dA inclined an angle of (90° – α) to the direction of flow. The drag from wall shear is τ_w sin αdA and the form pressure is P cos αdA. The total drag on the body is the sum of the integrals of the quantities each evaluated over the entire surface of the body in contact with fluid. The integrated drag from wall shear is called 'Wall Drag' and the total integrated drag from pressure is called 'Form Drag'.

For potential flow, $\tau_w = 0$ and there is no wall drag. Also, an equal force in the opposite direction balance the pressure drag in the direction of flow and the integral of the form drag zero. There is no net drag in potential flow. The phenomena causing both wall and form drag in actual fluids are complicated and cannot be calculated in general. For sphere and over regular shape at low fluid velocity the flow patterns and drag forces can be estimated by numerical methods. For irregular shapes and high velocities they must be determined by experiments.

4.3 DRAG CO-EFFICIENT

The drag co-efficient is a dimensionless quantity that is used to quantify the drag or resistance of an object in a fluid environment such as air or water. It is used in the drag equation in which a lower drag co-efficient indicates the object will have less aerodynamics or hydrodynamics drag. The drag co-efficient is always associated with a particular surface area.

Consider a smooth sphere immersed in a flowing fluid and at a distance from the solid boundary of the stream sufficient for the approaching stream to be in potential flow as shown in Fig. 4.3.

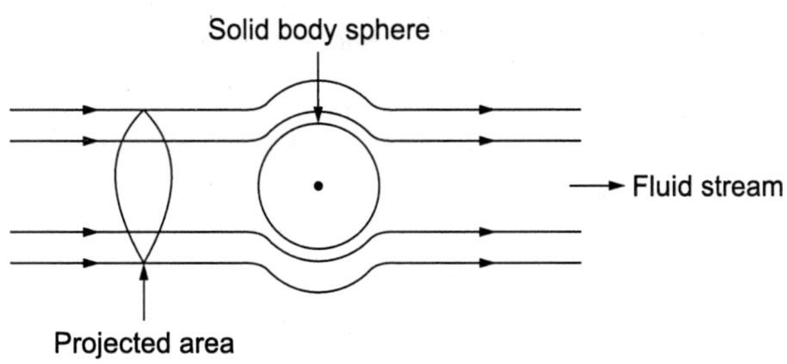

Fig. 4.3: Drag co-efficient

The projected area of the solid body is the area obtained by projecting the body on a plane perpendicular to the direction of flow. It is denoted by A_p.

4.4 PRESSURE DROP

The pressure drop depends on the average velocity in the channel $\bar{V}$ which is proportional to the superficial velocity or empty tower velocity and inversely proportional to the porosity. It may be expressed as

$$\bar{V} = \frac{V_0}{\varepsilon} \quad ...(4.1)$$

where, $\bar{V}$ = Average velocity

V_0 = Superficial velocity

ε = Porosity

With average velocity and channel size expressed in terms of measurable parameters $\bar{V}_0$, D_p and ε, the channel model can be used to predict the form of the correlation for pressure drop. For flow at very low Reynolds number the pressure should vary with first power of the velocity and inversely with the square of the channel size. The various empirical equations of pressure drops are discussed here.

4.4.1 Ergun Equation

The resistance to the flow of a fluid through the voids in a bed of solids is the resultant of the total drag of all the particles in the bed. Depending on the Reynolds number laminar flow, turbulent flow, form drag, separation and wake formation occurs. As in the drag of a single solid particle, there is no sharp transition between laminar and transition flow like that occurring in flow through conduits of constant cross-section.

Assumptions:

(*i*) It is assumed that the actual channels may be effectively replaced by a set of identical parallel conduit each of variable cross-section.

(*ii*) It is also assumed that the mean hydraulic radius of the channel is adequate to account for the variations in channel cross-sectional size and shape and the total drag per unit area of channel wall is the sum of two forces, viscous forces and inertial forces.

(*iii*) It is also assumed that particles are packed at random, with no preferred orientation of individual particles, that roughness are unimportant, that all particles are of the same size and shape and wall are negligible.

The assumption of additivity of viscous and inertial forces gives for F_D, the total force exerted in the fluid on the bed of solids. It may be expressed as

$$\frac{F_D}{A_S} = \frac{F_V}{A_S} + \frac{F_I}{A_S} \quad ...(4.2)$$

The viscous shear force per unit area of the channel wall can be expressed as

$$\frac{F_V}{A_S} = \frac{k_1 \propto V}{r_H} \quad ...(4.3)$$

where, $\frac{F_V}{A_S}$ = Viscous shear force per unit area of the channel

k_1 = Constant

μ = Viscosity of the fluid

V = Average velocity of the fluid

r_H = Hydraulic radius

At high velocity, inertial forces become large and viscous forces become negligible and viscosity is no longer a parameter. Hence, the inertial force per unit area can be expressed as

$$\frac{F_I}{A_S} = k_2 \rho V^2 \quad ...(4.4)$$

where, $\frac{F_I}{A_S}$ = Inertial force per unit area of the channel

k_2 = Constant

ρ = Density of the fluid

$\overline{V}$ = Average velocity of the fluid

Substituting equation (4.3) and equation (4.4) in equation (4.2), we get

$$\frac{F_D}{A_S} = \frac{k_1 \mu V}{r_H} + k_2 \rho V^2 \quad ...(4.5)$$

The velocity $\overline{V}$ is the average velocity through the channel in the bed. It is more convenient to use V_0, the velocity of the stream just before it encounters the layer of solids. This velocity is also called the superficial or empty tower velocity. It may be expressed as

$$\overline{V} = \frac{V_0}{\varepsilon} \quad ...(4.6)$$

where, $\overline{V}$ = Average velocity

V_0 = Superficial velocity

ε = Porosity

The total area of the channel boundaries is given by

$$A_S = N_P \cdot S_P \quad ...(4.7)$$

where, A_S = Total area of the channel boundaries

N_P = Number of particles in the bed

S_P = Area of one particle

The number of particles in the bed is given by the total volume of the solids in the tower divided by V_P the volume of one particle.

$$\Rightarrow \quad N_P = \frac{S_0 L(1-\varepsilon)}{V_P} \quad ...(4.8)$$

where, N_P = Number of particles in the bed

S_0 = Cross-sectional area of the bed

L = Length of the bed

V_P = Volume of one particle

Substituting equation (4.8) in equation (4.7), we get

$$A_S = \frac{S_0 L(1-\varepsilon) S_P}{V_P} \quad ...(4.9)$$

Hydraulic radius, r_H is defined as the ratio of the cross-section of the conduit to the perimeter of the conduit. If the numerator and denominator of this ratio are both multiplied by L, then r_H becomes the ratio of the volume of the voids in the tower to the total area of the solids.

$$\Rightarrow \quad r_H = \frac{S_0 L \varepsilon}{A_S} \quad ...(4.10)$$

$$\Rightarrow \quad r_H = \frac{S_0 L \varepsilon}{S_0 L(1-\varepsilon) S_P / V_P}$$

$$\Rightarrow \quad r_H = \frac{\varepsilon V_P}{(1-\varepsilon) S_P} \quad ...(4.11)$$

Substituting equations (4.6), (4.9) and (4.11) in equation (4.5), we get

$$\Rightarrow \qquad F_D = \frac{S_0 L(1-\varepsilon)S_P}{V_P}\left[\frac{k_1 \mu V_0 (1-\varepsilon)S_P}{\varepsilon^2 V_P} + \frac{k_2 \rho V_0^2}{\varepsilon^2}\right]$$

$$\Rightarrow \qquad F_D = \frac{S_0 L(1-\varepsilon)S_P}{\varepsilon^2 V_P}\left[\frac{k_1 \mu V_0 (1-\varepsilon)S_P}{V_P} + k_2 \rho V_0^2\right]$$

$$\Rightarrow \qquad F_D = \frac{S_0 L(1-\varepsilon)\rho S_P}{\varepsilon^2 V_P}\left[\frac{k_1 \mu V_0 (1-\varepsilon)S_P}{\rho V_P} + k_2 V_0^2\right] \qquad \text{...(4.12)}$$

The pressure drop over the bed ΔP may be used in place of the total force F_D by noting at this force must equal the product of the pressure drop and the cross-sectional area of the channel $S_o\varepsilon$.

$$\Rightarrow \qquad F_D = \Delta P S_o \varepsilon \qquad \text{...(4.13)}$$

Substituting equation (4.13) in equation (4.12), we get

$$\Delta P S_o \varepsilon = \frac{S_0 L(1-\varepsilon)\rho S_P}{\varepsilon^2 V_P}\left[\frac{k_1 \mu V_0 (1-\varepsilon)S_P}{\rho V_P} + k_2 V_0^2\right]$$

$$\Delta P \varepsilon = \frac{L(1-\varepsilon)\rho S_P}{\varepsilon^2 V_P}\left[\frac{k_1 \mu V_0 (1-\varepsilon)S_P}{\rho V_P} + k_2 V_0^2\right] \qquad \text{...(4.14)}$$

The sphericity ϕ_s is the ratio of the surface area of the sphere to the actual surface area of the particle.

$$\Rightarrow \qquad \phi_s = \frac{\text{Surface area of the sphere}}{\text{Actual surface area of the particle}}$$

$$\Rightarrow \qquad \phi_s = \frac{6V_P}{D_P S_P}$$

$$\Rightarrow \qquad \frac{S_P}{V_P} = \frac{6}{\phi_s D_P} \qquad \text{...(4.15)}$$

Substituting equation (4.15) in equation (4.14), we get

$$\Rightarrow \qquad \Delta P \varepsilon = \frac{L(1-\varepsilon)\rho 6}{\varepsilon^2 \phi_s D_P}\left[\frac{k_1 \mu V_0 (1-\varepsilon)6}{\rho \phi_s D_P} + k_2 V_0^2\right]$$

$$\Rightarrow \qquad \Delta P \varepsilon = \frac{L(1-\varepsilon)\rho V_0^2 6}{\varepsilon^2 \phi_s D_P}\left[\frac{k_1 \mu (1-\varepsilon)6}{\rho \phi_s D_P V_0} + k_2\right] \qquad \text{...(4.16)}$$

Ergan has correlated experimental data satisfactorily by considering the values of k_1 = 150/36 and k_2 = 1.75/6. Substituting the values of constant k_1 and k_2 in equation (4.16), we get

$$\Rightarrow \qquad \Delta P \varepsilon = \frac{L(1-\varepsilon)\rho V_0^2 6}{\varepsilon^2 \phi_s D_P}\left[\frac{150\mu (1-\varepsilon)6}{36\phi_s D_P V_0} + \frac{1.75}{6}\right]$$

$$\Rightarrow \qquad \frac{\Delta P \phi_s D_P \varepsilon^3}{\rho L V_0^2 (1-\varepsilon)} = \frac{150(1-\varepsilon)}{\phi_s (D_P V_0 \rho/\mu)} + 1.75 \qquad \text{...(4.17)}$$

This equation is called Ergan equation. It does not apply well to such extreme shape as long thin needles or the rings and saddle shaped solids are used as tower packing.

4.4.2 Kozeny-Carman Equation

It is the relation used in the fluid dynamics to calculate the pressure drop of a fluid flowing through a packed bed of solids. The equation was derived by Josef Kozeny and Philip C. Carman in 1927. The equation is valid for only Laminar flow.

From the Ergan equation we have

$$\frac{\Delta P \phi_s D_P \varepsilon^3}{\rho L V_0^2 (1-\varepsilon)} = \frac{150(1-\varepsilon)}{\phi_s (D_P V_0 \rho/\mu)} + 1.75 \quad \text{...(4.17)}$$

At laminar flow the second term of the right hand side of equation (4.17) is negligible. Hence, neglecting the second term, we get

$$\frac{\Delta P \phi_s D_P \varepsilon^3}{\rho L V_0^2 (1-\varepsilon)} = \frac{150(1-\varepsilon)}{\phi_s (D_P V_0 \rho/\mu)} \quad \text{...(4.18)}$$

$$\Rightarrow \quad \frac{\varnothing P}{L} = \frac{150(1-\varepsilon)\mu}{\phi_s D_P V_0 \rho} \times \frac{\rho V_0^2 (1-\varepsilon)}{\phi_s D_P \varepsilon^3}$$

$$\Rightarrow \quad \frac{\varnothing P}{L} = \frac{150\mu(1-\varepsilon)^2 V_0}{\phi_s^2 D_P^2 \varepsilon^3} \quad \text{...(4.19)}$$

where,
- ΔP = Pressure drop, N/m^2
- L = Length of the packed bed, m
- μ = Viscosity of the fluid, kg/m.s
- ε = Porosity
- V_0 = Superficial velocity, m/s
- ϕ_s = Sphericity
- D_P = Particle diameter, m

This equation is called Kozeny-Carman equation. For a given system, it indicates that the flow rate is directly proportional to the pressure drop. This statement is known as Darcy's law.

4.4.3 Blake Plummer Equation

For large Reynolds number the first term of the right hand side of equation (4.17) fadeout as viscous force is negligible and inertial force control. From the Ergan equation, we have

$$\frac{\Delta P \phi_s D_P \varepsilon^3}{\rho L V_0^2 (1-\varepsilon)} = \frac{150(1-\varepsilon)}{\phi_s (D_P V_0 \square/\mu)} + 1.75 \quad \text{...(4.17)}$$

Neglecting viscous force, we get

$$\frac{\Delta P \phi_s D_P \varepsilon^3}{\rho L V_0^2 (1-\varepsilon)} = 1.75$$

$$\Rightarrow \quad \frac{\varnothing P}{L} = \frac{1.75(1-\varepsilon)\rho V_0^2}{\phi_s D_P \varepsilon^3} \quad \text{...(4.20)}$$

This equation is called Blake Plummer equation.

4.5 FLUIDIZATION

Fluidization is a method to keep solid particles floating in an upward direction in a flow of fluid gas or liquid.

It consists of a fluidized chamber, provisions for fluid inlet, fluid outlet, emptying of fluid and a manometer. A typical fluidization bed is shown in Fig. 4.4.

When a fluid is passed at low velocity up through a bed of solid particles, the particles do not move and the pressure drop is given by the Ergan equation. If fluid velocity is steadily increased, the pressure drop and the drag on individual particle increases, and eventually the particle start to move and become suspended in the fluid. The fluidization and fluidized bed are used to describe the condition of fully suspended particles, since the suspension behave as a dense fluid. If the bed is fitted the top surface attains horizontal and large object will either float or sink in the bed depending on their density relative to the suspension. The fluidized solids can be drained from the bed through just like liquid. This fluidity is one of the main advantages of the use of fluidization for handling solids.

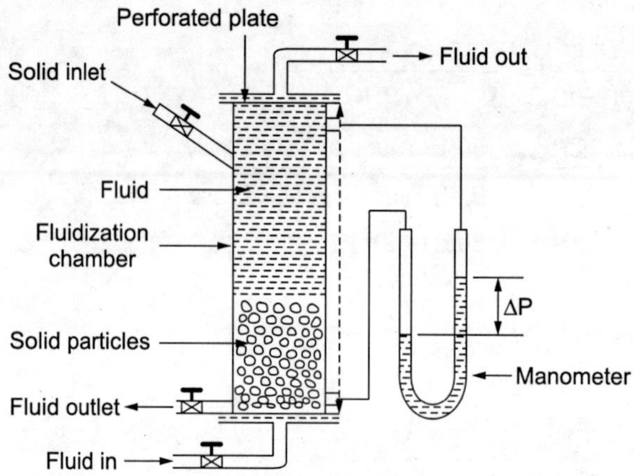

Fig. 4.4: Fluidized bed

4.5.1 Condition for Fluidization

Consider the fluidized bed partly filled with small granular material such as catalytic cracking catalyst. The bed may be open at the top and has a perforated plate at the bottom to support the bed of catalyst and to distribute the flow uniformly over the entire cross-section. Fluid admitted below the distributor plate at a low flow rate and pressure upward without causing any particle motion. If the particles are quite small, flow in the channels between the particles will be laminar and the pressure drop across the bed will be proportional to the superficial velocity V_0. As the velocity is gradually increased, the pressure drop increases but the particle do not move and the bed height remains the same. At certain velocity, the pressure drop across the bed counter balances the force of gravity on the particles or the weight of the bed and any further increase in velocity causes the particle to move. Sometime the bed expands slightly with the grains still in contact, since just a slight increase in porosity ε can offset an increase of several percent in superficial velocity and keep ΔP constant. With a further increase in velocity, the particles becomes separated enough to move about in the bed and true fluidizations begins.

Once the bed is fluidized, the pressure drop across the bed stays constant but the bed height continues with increasing flow. The bed can be operated at quite high velocity with very little or no loss of solids. Since the superficial velocity needed to support a bed of particles is much less than the terminal settling velocity for individual particles. The effect of pressure on superficial velocity is shown in Fig. 4.5.

If the flow rate to the fluidized bed is gradually reduced, the pressure drop remains constant and the bed height decreases. However, the final bed height may be greater than the initial due to the fixed bed. Since, solid dumped in the chamber tend to pack more tightly than solids newly settling from fluidized state. The pressure drop at low velocity is less than that of the original fixed bed. On starting up again the pressure drop offset the weight of the bed. At this point should be considered to give the minimum fluidization velocity V_m. To increase, V_m the be should be fluidized vigorously and allowed to settle with the fluid opened off. And the low rate increased gradually until the bed start to expands. More producible value of V_m can sometime be obtained from the intersection of the graph ΔP the fixed bed and the fluidized bed.

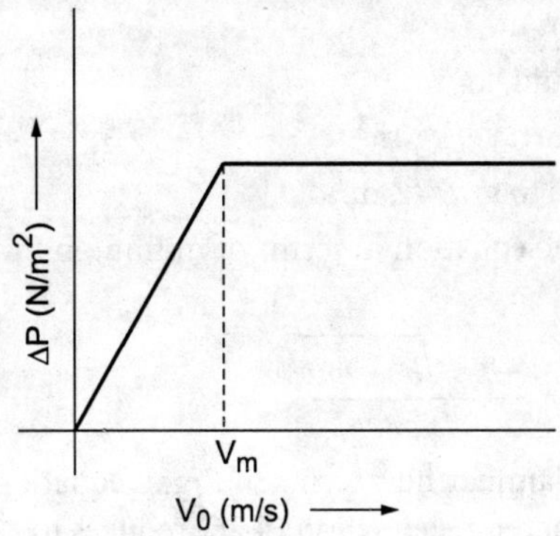

Fig. 4.5: Effect of pressure on superficial velocity

4.5.2 Minimum Fluidization Velocity

An equation for the minimum fluidization velocity can be obtained by setting the pressure drop across the bed equal to the weight of the bed per unit area of cross-section, allowing for the buoyant force for the displaced fluid. It can be expressed as

$$\frac{\varnothing P}{L} = g(1 - \varepsilon)(\rho_P - \rho) \quad ...(4.21)$$

At incipient fluidization the porosity is minimum ε_m. Thus, equation (4.21) can be written as

$$\Delta P = g(\rho_P - \rho)(1 - \varepsilon_m)L \quad ...(4.22)$$

The pressure drop in packed bed expressed by Ergan equation as

$$\frac{\Delta P \phi_s D_P \varepsilon^3}{\rho L V_0^2 (1-\varepsilon)} = \frac{150(1-\varepsilon)\mu}{\phi_s D_P V_0 \rho} + 1.75 \quad ...(4.17)$$

$$\Rightarrow \quad \frac{\varnothing P}{L} = \frac{150(1-\varepsilon)^2 \mu V_0}{\phi_s^2 D_P^2 \varepsilon^3} + \frac{1.75(1-\varepsilon)\rho V_0^2}{\phi_s D_P \varepsilon^3} \quad ...(4.23)$$

Applying to the point of incipient fluidization, we get

$$\Delta P = \frac{150(1-\varepsilon_m)^2 \mu V_m L}{\phi_s^2 D_P^2 \varepsilon_m^3} + \frac{1.75(1-\varepsilon_m)\rho V_m^2 L}{\phi_s D_P \varepsilon_m^3} \quad ...(4.24)$$

Comparing equation (4.22) and equation (4.24), we get

$$\frac{150(1-\varepsilon_m)^2 \mu V_m L}{\phi_s^2 D_P^2 \varepsilon_m^3} + \frac{1.75(1-\varepsilon_m)\rho V_m^2 L}{\phi_s D_P \varepsilon_m^3} = g(\rho_P - \rho)(1 - \varepsilon_m)L \quad ...(4.25)$$

$$\Rightarrow \quad \frac{150(1-\varepsilon_m)\mu V_m L}{\phi_s^2 D_P^2 \varepsilon_m^3} + \frac{1.75 \rho V_m^2 L}{\phi_s D_P \varepsilon_m^3} = g(\rho_P - \rho)L \quad ...(4.26)$$

where, ε_m = Minimum porosity
V_m = Minimum fluidization velocity, m/s
μ = Viscosity of the fluid, kg/m.s
L = Bed length
ϕ_s = Sphericity
D_P = Particle diameter, m
ρ = Density of the fluid, kg/m^3
ρ_P = Density of the particle, kg/m^3
g = Acceleration due to gravity, m/s^2

The equation (4.26) is a quadratic equation in terms of minimum fluidization velocity V_m. Hence, it can be solve using the formula.

$$x = \frac{-b \pm \sqrt{b^2 - 4ac}}{2a} \quad ...(4.27)$$

For very small particle only the laminar flow terms of Ergan equation is significant and the term related to inertial forces can be neglected. Therefore, equation (4.26) reduces to

$$\frac{150(1-\varepsilon_m)\mu V_m L}{\phi_s^2 D_P^2 \varepsilon_m^3} = g(\rho_P - \rho)L \quad ...(4.28)$$

$$\Rightarrow \quad V_m = \frac{g(\rho_P - \rho)\varepsilon_m^2 \phi_s^2 D_P^2}{150(1-\varepsilon_m)\mu} \quad ...(4.29)$$

This the modified Ergan equation can be used to calculate minimum fluidization velocity.

4.5.3 Types of Fluidization

The fluidization may be classified into following types

(*i*) Particulate fluidization
(*ii*) Aggregate fluidization
(*iii*) Bubbling fluidization
(*iv*) Slugging fluidization
(*v*) Turbulent fluidization
(*vi*) Circulating fluidization
(*vii*) Continuous fluidization

4.5.3.1 Particulate Fluidization

The equation derived for the minimum fluidization velocity is applicable for liquid as well as gases, but beyond minimum fluidization velocity the appearance of bed fluidized with liquids or gases is often quite different.

For example, when fluidizing sand with water, the particle move further apart and their motion become more vigorous as the velocity is increased. But the average bed density at given velocity is same in al section of the bed. This is called particulate fluidization. It is characterized by large but uniform expansion of the bed at high velocity. For particulate fluidization, we have

$$\frac{\varnothing P}{L} = \frac{150(1-\varepsilon)^2 \mu V_0}{\phi_s^2 D_P^2 \varepsilon^3} \quad ...(4.30)$$

From the incipient fluidization, we also have

$$\frac{\varnothing P}{L} = g(1-\varepsilon)(\rho_P - \rho) \quad ...(4.21)$$

$$\Rightarrow \quad g(1-\varepsilon)(\rho_P - \rho) = \frac{150(1-\varepsilon)^2 \mu V_0}{\phi_s^2 D_P^2 \varepsilon^3}$$

$$\Rightarrow \quad \frac{\varepsilon^3}{(1-\varepsilon)} = \frac{150\mu V_0}{g\phi_s^2 D_P^2(\rho_P - \rho)} \quad ...(4.31)$$

$$\Rightarrow \quad V_0 = 10\, V_m \quad ...(4.32)$$

4.5.3.2 Aggregate Fluidization

At superficial velocity somewhat greater than minimum fluidization velocity, most of the gas passes through the bed as bubbles or voids, which are almost free of solids and only a small fraction of the gas flows in the channels between the particles. The particles move erratically and supported by the fluid, but in the space between bubbles, the void fraction is about the same as that at incipient fluidization. The non-uniform nature of the bed is at firs attributed to aggregation of the particles during fluidization. This phenomena is called Aggregate fluidization.

4.5.3.3 Bubbling Fluidization

In aggregate fluidization there was no evidence that the particles stick together. Hence, this phenomena is called Bubbling fluidization. The bubbles that form behave as air bubbles in water or bubbles vapour in boiling fluid. Studies that account for the rate of heat transfer or chemical reaction in a bubbling bed refer to the two phase theory of fluidization. In this bubbles are one phase and dense bed of suspended particles in the second phase. It may be expressed as

$$\Delta P = f_b \cdot u_b + (1 - f_b)V_m \quad ...(4.33)$$

4.5.3.4 Slugging Fluidization

The behavior of bubbling fluidization depends very strongly on the number and size of the gas bubbles, which are often hard to predict. The average bubble size depends on the nature and size distribution of the particles, the type of distributor plate, the superficial velocity and the depth of the bed. Bubbles tends to coalesce and grow as they rise through the fluidized bed and maximum stable bubble size may be several inches to a few feet in diameter. If a small diameter column is used with a deep bed solids, the bubble may grow until they fill the entire cross-section. Successive bubbles can travel up the column separated by slugs of solids. It is undesirable because of pressure fluctuation in the bed.

4.5.3.5 Turbulent Fluidization

When the superficial gas velocity increases to values much above V_m, there is transition from bubbling fluidization to what is called Turbulent fluidization. It is also called fast fluidization. The transition occurs when the bed is expanded so much that there can be no longer small options of high or low density, with rapid density fluctuations at all points in the bed. The velocity for transition to turbulent fluidization is generally in the range of 0.3 to 0.6 m/s. It is difficult to predict the transition velocity because it depends on the particles properties and the average bubble size.

4.5.3.6 Circulating Fluidization

At still higher gas velocities, all particles fed are rapidly trained with the gas but can be recovered with a cyclone and returned to the bottom to maintain solids in the unit. This type of system is called circulating fluidization. Even though there is no distinct bed of solids. The volume of solids in the suspension is quite low. Such system is quite like an example of pneumatic transport.

4.5.3.7 Continuous Fluidization

When the velocity of the fluid through the solids becomes large enough, all the particles are entrained In the fluid and carried along with it to give continuous fluidization. Some gas solids reactions operate in this fraction, but principal applications of continuous fluidization are in transporting solids from point to point.

4.5.4 Application of Fluidization

Extensive use of fluidization began in the petroleum industry with the development of fluidized bed catalytic cracking, although the industry now generally uses riser or transport line reactors for catalytic cracking rather than fluidized bed. The catalyst regeneration is still carried out in fluidized bed reactors, which is as large as 10 inch in diameter. Fluidization is used in other catalytic processes such as the synthesis of acrylonitrile and for carried out solids gas reactions.

There is much interest in the fluidized bed combustion of coal as a means of reducing boiler cost and decreasing the emission of pollutants. Fluidized bed is also used for roasting ores, drying fine solids and adsorption of gases.

4.5.5 Slurry Transport

Particles smaller than about 50 μm in diameter settle very slowly and are readily suspended and when the diameter is 0.25 mm or greater a fairly large liquid velocity is needed to keep the particles from moving at all, especially in horizontal pipe. The critical velocity below which particles will settle out in typically between 1 and 5 m/s, depending on the density difference between solids and liquids, the particle diameter, the slurry concentration and the size of the chamber. Critical velocities are larger in big chamber than in small chamber.

The pressure drop in slurries of non settling particles may be found from the equations for a homogeneous liquid with appropriate allowance for the increases density and apparent velocity. For settling slurries there is no single satisfactory correlation. The pressure drop in a horizontal pipe is greater than that in a single phase fluid of the same density and viscosity as the slurry especially near the critical velocity, but approaches that in the single phase liquid as the velocity increases. The velocity in a long slurry pipeline is typically 1.5 to 2 times of V_c.

4.5.5.1 Pneumatic Conveying

When a high velocity of gas is passed through a bed of solids particles, the individual particles are dragged with the air, finally they suspended in the air. The suspended fluid in a pneumatic conveyor is a gas usually air, flowing at velocities between 1.5 to 30 m/s in a pipe ranging from 50 to 400 mm in diameter.

The most important material handling techniques in the chemical industry is the movement of material suspended in a stream of air over horizontal and vertical distances ranging from a few to several hundred feet. Materials ranging from fine powders to more in density 3200 kg/m^3 can be handled. The capacity of a pneumatic conveying system depends on particles bulk density, particle size and shape to some extent, energy content of the conveying air over the entire system, diameter of conveying line and equivalent length of conveying line. A typical pneumatic conveyor is shown in Fig. 4.6.

Minimum capacity is achieved when the energy of conveying air is just sufficient to move the product through the line without stoppage. To prevent such stoppage, it is good practice to provide an additional increment of air energy so that a factor of safety exists that allows air a factor for minor changes in product characteristics. An optimum system is one that always through operating economy, all design factors above the minimum required. Generally, pneumatic conveyors are classified according to the basic types, pressure, vacuum, combination of pressure and vacuum, fluidization and blow tank.

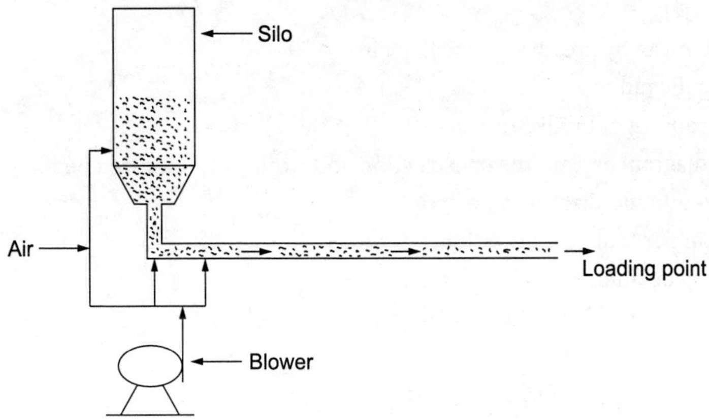

Fig. 4.6: Pneumatic conveyor

Fluidizing system generally convey pre-fluidized, finely divided, non free flowing materials for short distance such as from storage bin or transportation vehicles to the entrance of a bin conveying system. A particular advantage storage bin applications is that the bottom of bin is permitted to the nearly horizontal. Fluidizing is accomplished by means of a chamber in which air is pass through a porous membrane that forms the bottom of the conveyor, upon which the material to be conveyed rests. As air passes through the membrane, a film of air surrounds each particle. At the point of incipient fluidization the material takes on the characteristics of free flow. It can then be passed into conveyor air stream by a rotary feeder.

It is used in the formulation of powdered material insuffiations of light and bulky materials like grains can be transmitted. It is also used to transportation of unpleasant injection material, since it is dual system.

REVIEW QUESTIONS

4.1 Define drag.
4.2 What do you mean by drag co-efficient? Explain.
4.3 Define form drag.
4.4 Define wall drag.
4.5 Define total drag.
4.6 What do you mean by pressure drop? Explain.
4.7 Define hydraulic radius.
4.8 Explain the concept of fluidization.
4.9 With a neat diagram explain the construction and working of fluidized bed.
4.10 What is minimum fluidization velocity?
4.11 Briefly explain particulate fluidization.
4.12 What is aggregate fluidization? Explain.
4.13 Explain the concept of Bubbling fluidization.
4.14 What is slugging fluidization? Explain.
4.15 Explain the concept of turbulent fluidization.
4.16 Briefly explain circulating fluidization.
4.17 What is circulating fluidization? Explain.
4.18 Explain the application of fluidization.
4.19 Write briefly on slurry transport.
4.20 With a neat diagram explain the construction and working of pneumatic conveying.

EXERCISES

4.1 Starting from the basic derive Ergan equation.
4.2 Starting from the basic derive Kozeny Carmon equation.
4.3 Starting from the basic derive Blake Plummer equation.
4.4 Derive an equation for minimum fluidization velocity.

❑❑❑

5 CHAPTER Motion of Particles Through Fluids

5.1 INTRODUCTION

The particles are quite, closed together and move with random motion in liquid throughout the container. Particles move rapidly in all directions but collide with each other more frequently than in gases due to shorter distances between the particles. The separation of particles in mechanical operation involves the movement of the solid particles through a fluid. for example, elimination of dust and fumes, removal of solids from liquids waste to allow discharges into public drainage system.

5.2 MECHANISMS OF PARTICLE MOTION

The movement of particle through a fluid requires external force acting on the particle. This force may come from a density difference between particle and the fluid. Here, gravitational or centrifugal force is considered. The forces acting on a particle moving through fluid are:

(*i*) External force

(*ii*) Buoyant force

(*iii*) Drag force

5.2.1 External Force

It is the gravitational or centrifugal force applied on the particle.

5.2.2 Buoyant Force

It acts parallel with the external force but in the opposite direction.

5.2.3 Drag Force

It appears whenever there is relative motion between the particle and the fluid.

One dimensional motion of particle through fluid: Consider a particle of mass m moving through a fluid under the action of an external force F_e. Let the velocity of the particle relative to the fluid be u. Let the buoyant force on the particle be F_b. Let the drag force be F_D. Then from the force balance, we have

$$\frac{m}{g_c}\frac{du}{dt} = F_e - F_b - F_D \quad \text{...(5.1)}$$

where, g_c = Newton's law proportionality factor

$\frac{du}{dt}$ = Acceleration of the particle

The external force can be expressed as a product of the mass and the acceleration of the particle from the external force. It may be expressed as

$$F_e = \frac{ma_e}{g_c} \quad ...(5.2)$$

By Archimedes principle the buoyant force is the product of the mass of the fluid displaced by the particle and the acceleration from the external force. It may be expressed as

$$F_b = \frac{m\rho a_e}{\rho_P g_c} \quad ...(5.3)$$

where, ρ = Density of the fluid

ρ_P = Density of the particle

a_e = Acceleration of the particle from the external force

The drag force can be expressed as

$$F_D = \frac{C_D u^2 \rho A_P}{2g_c} \quad ...(5.4)$$

where, C_D = Dimensionless drag coefficient

A_P = Projected area of the particle

u = Velocity of the fluid

Substituting equation (5.2), (5.3) and (5.4) in equation (5.1), we get

$$\Rightarrow \quad \frac{m}{g_c}\frac{du}{dt} = \frac{ma_e}{g_c} - \frac{m\rho a_e}{\rho_P g_c} - \frac{C_D u^2 \rho A_P}{2g_c} \quad ...(5.5)$$

$$\Rightarrow \quad \frac{du}{dt} = a_e - \frac{\rho a_e}{\rho_P} - \frac{C_D u^2 \rho A_P}{2m}$$

$$\Rightarrow \quad \frac{du}{dt} = \frac{a_e(\rho_P - \rho)}{\rho_P} - \frac{C_D u^2 \rho A_P}{2m} \quad ...(5.6)$$

Motion from gravitational force: If the external force is gravity a_e is g the acceleration due to gravity. Then, the equation (5.6) is reduced to

$$\frac{du}{dt} = \frac{g(\rho_P - \rho)}{\rho_P} - \frac{C_D u^2 \rho A_P}{2m} \quad ...(5.7)$$

Motion in centrifugal field: A centrifugal force appears whenever the direction of movement of a particle is changed. The acceleration from a centrifugal force for a circular motion may be expressed as

$$a_e = r\omega^2 \quad ...(5.8)$$

where, r = Radius of the path of particle

ω = Angular velocity

Substituting equation (5.8) in equation (5.6), we get

$$\frac{du}{dt} = \frac{r\omega^2(\rho_P - \rho)}{\rho_P} - \frac{C_D u^2 \rho A_P}{2m} \quad ...(5.9)$$

In this equation u is the velocity of the particle relative to the fluid and is moving outwardly along with radius.

5.3 TERMINAL VELOCITY

In gravitational settling g is constant. Also, the drag always increases with velocity. The acceleration of particle decreases with time and approaches to zero. The particle quickly reaches a constant velocity, which is the maximum attainable under the circumstances. The velocity attained by the particle at this condition is called terminal velocity.

The equation for the terminal velocity u_t, may be obtained for gravitational settling by taking $du/dt = 0$ and substituting in equation (5.6), we get

$$0 = \frac{g(\rho_P - \rho)}{\rho_P} - \frac{C_D u_t^2 \rho A_P}{2m} \quad \text{...(5.10)}$$

$$\Rightarrow \quad \frac{C_D u_t^2 \rho A_P}{2m} = \frac{g(\rho_P - \rho)}{\rho_P}$$

$$\Rightarrow \quad u_t^2 = \frac{2g(\rho_P - \rho)m}{\rho A_P C_D \rho_P}$$

$$\Rightarrow \quad u_t = \sqrt{\frac{2g(\rho_P - \rho)m}{\rho A_P C_D \rho_P}} \quad \text{...(5.11)}$$

In motion form a centrifugal force, the velocity depends on the radius and acceleration is not constant if the particle is in motion with respect to the fluid.

In many practical uses of centrifugal force, however du/dt is small in comparison with the other terms in equation (5.9). Hence, du/dt may be neglected in equation (5.9) to get terminal velocity at any given radius.

$$\Rightarrow \quad 0 = \frac{r\omega^2(\rho_P - \rho)}{\rho_P} - \frac{C_D u_t^2 \rho A_P}{2m} \quad \text{...(5.12)}$$

$$\Rightarrow \quad \frac{C_D u_t^2 \rho A_P}{2m} = \frac{r\omega^2(\rho_P - \rho)}{\rho_P}$$

$$\Rightarrow \quad u_t^2 = \frac{2mr\omega^2(\rho_P - \rho)}{C_D \rho A_P \rho_P}$$

$$\Rightarrow \quad u_t = \omega \times \sqrt{\frac{2mr(\rho_P - \rho)}{C_D \rho A_P \rho_P}} \quad \text{...(5.13)}$$

5.4 DRAG CO-EFFICIENT

The drag co-efficient C_D for various substances as a function of Reynolds number is shown in Fig. 5.1.

The curve for spherical crushed limestone is described here. These are, however apply only under restricted conditions. The particle must be solids. It must be far from the other particles and from the vessel walls so that the flow pattern around the particle is not distorted and it must be moving at its terminal velocity with respect to the fluid. The drag co-efficient for accelerating particles are appreciably greater, so a particle dropped in a still fluid take larger time to reach terminal velocity than would be predicted using the steady state values of C_D. Particles injected into fast flowing stream also accelerate more slowly than expected and the drag co-efficient in the case are therefore less than the normal values. However, for most processes involving small particle or drops, the time for acceleration to the terminal velocity is still quite small and is often ignored in analysis of the process.

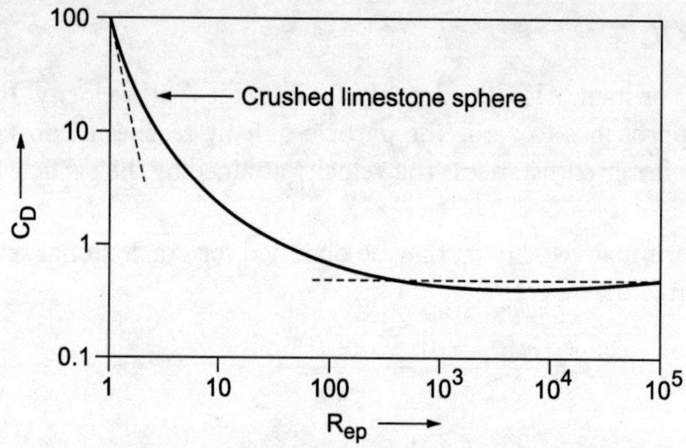

Fig. 5.1: Drag co-efficient for sphere

The variation in particle shape can be accounted for by obtaining separate curves of C_D V_S R_{ep} for each shape. In the free motion of non-spherical particle through a fluid, the orientation is constantly changing. This change consume energy, increasing the effective drag on the particle and C_D greater than for the motion of fluid past a fixed particle. For example, the drag co-efficient of crushed limestone sphere.

5.5 MOTION OF SPHERICAL PARTICLE

Consider the motion of spherical particle in a liquid in rest as shown in Fig. 5.2.

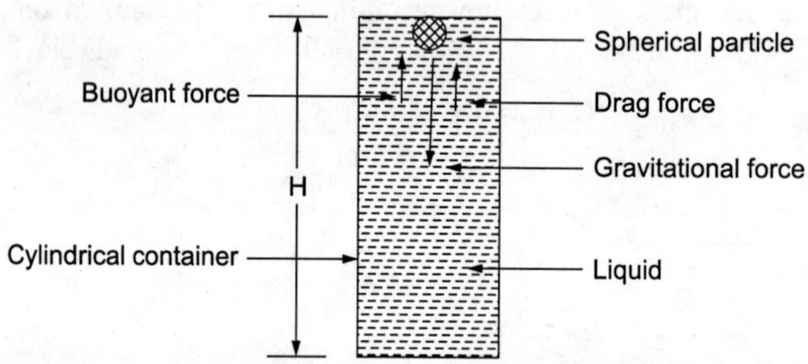

Fig. 5.2: Motion of spherical particle

Under this condition, the gravitational force of the particle acts downward where as the buoyant force and drag force acts on the opposite direction. The force balance may be expressed as

$$F_g = F_b + F_D \quad \text{...(5.14)}$$

where, F_g = Gravitational force

F_b = Buoyant force

F_D = Drag force

The gravitational force offered by the body may be expressed as

$$F_g = \frac{\pi}{6} d_P^3 \rho_P g \quad \text{...(5.15)}$$

where, d_P = Particle diameter

ρ_P = Density of the particle

g = Acceleration due to gravity

The buoyant force offered by the liquid on the body may be expressed as

$$F_b = \frac{\pi}{6} d_P^3 \rho g \quad ...(5.16)$$

where, ρ = Density of the liquid

The drag force offered by the liquid on the body may be expressed as

$$F_D = \phi(N_{Rep}) \rho u_t^2 \frac{\pi d_P^2}{4} \quad ...(5.17)$$

where, ρ = Density of the liquid

u_t = Terminal velocity

Substituting equation (5.15), (5.16) and (5.17) in equation (5.14), we get

$$\frac{\pi}{6} d_P^3 \rho_P g = \frac{\pi}{6} d_P^3 \rho g + \phi(N_{Rep}) \rho u_t^2 \frac{\pi}{4} d_P^2 \quad ...(5.18)$$

$$\Rightarrow \quad \frac{\pi}{6} d_P^3 (\rho_P - \rho) g = \phi(N_{Rep}) \rho u_t^2 \frac{\pi}{4} d_P^2 \quad ...(5.19)$$

For motion of the particle in laminar region, the particle Reynolds number may be expressed as

$$\phi(N_{Rep}) = \frac{12\mu}{d_P u_t \rho} \quad ...(5.20)$$

where, μ = Viscosity of the liquid

Substituting equation (5.20) in equation (5.19), we get

$$\frac{\pi}{6} d_P^3 (\rho_P - \rho) g = \frac{12\mu}{d_P u_t \rho} \times \rho u_t^2 \frac{\pi}{4} d_P^2 \quad ...(5.21)$$

$$\Rightarrow \quad d_P^2 (\rho_P - \rho) g = 18 \mu u_t$$

$$\Rightarrow \quad u_t = \frac{d_P^2 (\rho_P - \rho) g}{18\mu} \quad ...(5.22)$$

where, u_t = Terminal settling velocity, m/s

d_p = Diameter of spherical particle, m

ρ_P = Density of the particle, kg/m^3

ρ = Density of the liquid, kg/m^3

μ = Viscosity of the liquid, kg/m.s

g = Acceleration due to gravity, m/s^2

This is the equation of motion of spherical particle in laminar flow region where Reynolds number is less than 2. This is also called Stokes law and equation for terminal settling velocity.

In the case of turbulent and transition region, the function of N_{Rep} is not a simple function as in the case of laminar flow. This function is replaced by the term called drag co-efficient. Drag force in turbulent and transition region may be obtained by using drag co-efficient C_D which is equal to 2 N_{Rep} in laminar flow.

For turbulent flow:

$$2 N_{Rep} = C_D \quad ...(5.23)$$

$$\Rightarrow \quad N_{Rep} = \frac{C_D}{2} \quad ...(5.24)$$

Substituting equation (5.24) in equation (5.18), we get

$$\frac{\pi}{2}d_P^3(\rho_P-\rho)g = \frac{C_D}{2}\rho u_t^2\frac{\pi}{4}d_P^2 \quad ...(5.25)$$

$$\Rightarrow \quad u_t^2 = \frac{4d_P(\rho_P-\rho)g}{3\rho C_D}$$

$$\Rightarrow \quad u_t = \left[\frac{4d_P(\rho_P-\rho)g}{3\rho C_D}\right]^{0.5} \quad ...(5.26)$$

For Stokes law: $u_t = k_1D^2$ ...(5.27)

In turbulent region C_D is almost constant

$$u_t = k_2[D]^{0.5} \quad ...(5.28)$$

In transition region C_D varies

$$u_t = k_3\left[\frac{D}{C_D}\right]^{0.5} \quad ...(5.29)$$

For 1000 < N_{Rep} < 2,00,000 the drag co-efficient is approximately constant and the equations are:

$$C_D = 0.44 \quad ...(5.30)$$

$$F_D = 0.055\,\pi d_p^2u_t^2\rho \quad ...(5.31)$$

$$u_t = 1.75\sqrt{\frac{d_P(\rho_P-\rho)g}{\rho}} \quad ...(5.32)$$

This equation is known as Newtons law and is applies only for fairly large particles falling in gases or low viscosity fluids.

The terminal velocity u_t varies with d_p^2 in the Stokes law range, whereas in the Newton's law range it varies with $d_p^{0.5}$.

Example 5.1. What will be the settling velocity of spherical steel ball having 0.38 mm diameter settling in an oil of specific gravity 0.82 and viscosity 0.1 poise. Assume the density of the steel ball 7.87 g/cm³.

Solution: From the terminal settling velocity, we have

$$u_t = \frac{d_P^2(\rho_P-\rho)g}{18\mu}$$

Data given:

$$d_P = 0.38\times10^{-3}\text{ m}$$

$$\rho_P = 7870\text{ kg/m}^3$$

$$\rho = 820\text{ kg/m}^3$$

$$\mu = 0.01\text{ kg/m.s}$$

$$g = 9.81\text{ m/s}^2$$

$$\Rightarrow \quad u_t = \frac{(0.38\times10^{-3})^2\times(7870-820)\times9.81}{18\times0.01}$$

$$= 0.055\text{ m/s}$$ **Ans.**

Example 5.2. Calculate the diameter of steel ball settling in an oil having the settling velocity 0.055 m/s. The density of the oil and steel ball are 820 kg/m³ and 7870 kg/m³ respectively. Assume the viscosity of oil 0.01 kg/m.s.

Solution: From the Stokes law, we have

$$u_t = \frac{d_P^2(\rho_P - \rho)g}{18\mu}$$

Data given:

$$u_t = 0.055 \text{ m/s}$$
$$\rho_P = 7870 \text{ kg/m}^3$$
$$\rho = 820 \text{ kg/m}^3$$
$$\mu = 0.01 \text{ kg/m.s}$$
$$g = 9.81 \text{ m/s}^2$$

$$\Rightarrow \quad 0.055 = \frac{d_P^2(7870 - 820) \times 9.81}{18 \times 0.01}$$

$$\Rightarrow \quad d_P^2 = 0.4131 \times 10^{-6}$$

$$\Rightarrow \quad d_P = 0.000378 \text{ m}$$
$$= 0.378 \text{ mm}$$
$$= 0.38 \text{ mm}$$ **Ans.**

Example 5.3. If the settling velocity of a spherical steel ball settling in an oil having density 820 kg/m^3 is 0.055 m/s. Calculate the density of the steel ball. The diameter of the steel ball and viscosity of oil are 0.38 mm and 0.01 kg/m.s respectively.

Solution: From the Stokes law, we have

$$u_t = \frac{d_P^2(\rho_P - \rho)g}{18\mu}$$

Data given:

$$u_t = 0.055 \text{ m/s}$$
$$d_P = 0.38 \text{ mm}$$
$$\rho = 820 \text{ kg/m}^3$$
$$\mu = 0.01 \text{ kg/m.s}$$
$$g = 9.81 \text{ m/s}^2$$

$$\Rightarrow \quad 0.055 = \frac{(0.38 \times 10^{-3})^2 \times (\rho_P - 820) \times 9.81}{18 \times 0.01}$$

$$\Rightarrow \quad (\rho_P - 820) = 6989$$

$$\Rightarrow \quad \rho_P = 7809 \text{ kg/m}^3$$ **Ans.**

Example 5.4. When the settling velocity of a spherical steel ball settling in an oil having density 820 kg/m^3 is 0.055 m/s. Calculate the viscosity of the oil in which steel ball is falling. The diameter and density of the steel ball are 0.38 mm and 7870 kg/m^3 respectively.

Solution: From the Stokes law, we have

$$u_t = \frac{d_P^2(\rho_P - \rho)g}{18\mu}$$

Data given:

$$u_t = 0.055 \text{ m/s}$$
$$d_P = 0.38 \text{ mm}$$
$$\rho_P = 7870 \text{ kg/m}^3$$
$$\rho = 820 \text{ kg/m}^3$$

$$g = 9.81 \text{ m/s}$$

$$\Rightarrow \quad 0.055 = \frac{(0.38\times10^{-3})^2\times(7870-820)\times9.81}{18\times\mu}$$

$$\Rightarrow \quad \mu = 0.01 \text{ kg/m.s} \quad \textbf{Ans.}$$

Example 5.5. What will be the settling velocity of a spherical particle of steel ball 0.4 mm diameter in an oil of specific gravity 0.82 and viscosity 0.001 Ns/m^2. The specific gravity of steel is 7.87. **[VTU Exam, May 2010]**

Solution: From the Stokes law, we have

$$u_t = \frac{d_P^2(\rho_P-\rho)g}{18\mu}$$

Data given:

$$d_P = 0.4 \text{ mm}$$
$$\rho_P = 7870 \text{ kg/m}^3$$
$$\rho = 820 \text{ kg/m}^3$$
$$\mu = 0.001 \text{ kg/m.s}$$
$$g = 9.81 \text{ m/s}^2$$

$$\Rightarrow \quad u_t = \frac{(0.4\times10^{-3})^2\times(7870-820)\times9.81}{18\times0.001}$$

$$\Rightarrow \quad u_t = 0.615 \text{ m/s} \quad \textbf{Ans.}$$

Example 5.6. Drops of oil 15 micrometer in diameter are to be settled from their mixture with air. The specific gravity of the oil is 0.90 and the air is at 21°C and 1 atm. A settling time of 1 minute is available. How high should be the chamber to allow settling of these particles?

Solution:

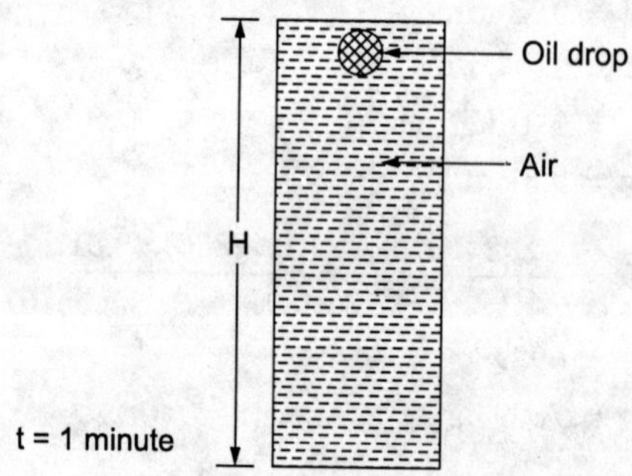

Fig. 5.3: Settling chamber

From the Stokes law, we have

$$u_t = \frac{d_P^2(\rho_p-\rho)g}{18\mu}$$

Data given:

$$d_P = 15 \times 10^{-6} \text{ m}$$
$$\rho_P = 900 \text{ kg/m}^3$$
$$\rho = 1.2 \text{ kg/m}^3 \text{ (From the data book)}$$
$$\mu = 1.81 \times 10^{-5} \text{ kg/m.s (From the data book)}$$
$$g = 9.81 \text{ m/s}$$

$$\Rightarrow \qquad u_t = \frac{(15\times10^{-6})^2\times(900-1.2)\times9.81}{18\times(1.81\times10^{-5})}$$

$$\Rightarrow \qquad u_t = 0.0061 \text{ m/s}$$

From the settling velocity, we have

$$u_t = \frac{H}{t} \qquad \text{...(5.33)}$$

where, u_t = Settling velocity, m/s

H = Height of the chamber, m

$\Rightarrow$ t = Time, m

$$\Rightarrow \qquad 0.0061 = \frac{H}{60}$$

$$\Rightarrow \qquad H = 0.366 \quad \textbf{Ans.}$$

Example 5.7. What will be the settling velocity of glass spherical particles having 0.25 mm diameter settling in water at 20°C. The density of glass sphere is 2450 kg/m^3.

Solution: From the Stokes law, we have

$$u_t = \frac{d_P^2(\rho_P-\rho)g}{18\mu}$$

Data given:

$$d_p = 0.25 \text{ mm}$$

$$\rho_p = 2450 \text{ kg/m}^3$$

$$\rho = 1000 \text{ kg/m}^3$$

$$\mu = 0.001 \text{ kg/m.s}$$

$$g = 9.81 \text{ m/s}^2$$

$$\Rightarrow \qquad u_t = \frac{(0.025\times10^{-3})^2\times(2450-1000)\times9.81}{18\times0.001}$$

$$\Rightarrow \qquad u_t = 0.049 \text{ m/s} \quad \textbf{Ans.}$$

5.6 CRITERIA FOR SETTLING OF PARTICLE

To identify the rage in which the motion of the particle lies, the velocity term is eliminated from the Reynolds number by substituting u_t for the Stokes law range, we get

$$N_{Rep} = \frac{D_P u_t \rho}{\mu} = \frac{D_P^3\rho(\rho_P-\rho)g}{18\mu^2} \qquad \text{...(5.34)}$$

If Stokes law is to apply N_{Rep} must be less than 1. To provide a convenient criterion K

Let

$$K = D_P\left[\frac{\rho(\rho_P-\rho)g}{\mu^2}\right]^{1/3} \qquad \text{...(5.35)}$$

Thus, from equation (5.34) we get

$$N_{Rep} = \frac{1}{18}K^3 \qquad \text{...(5.36)}$$

Setting N_{Rep}, equal to 1 and solving equation (5.36), we get

$$K = (18)^{1/3}$$

$$\Rightarrow \qquad K = 68.9$$

Thus, if K is greater than 68.9 and less than 2360, Newton's law applies. When K is greater than 2360, the drag co-efficient may change abruptly with small changes in fluid velocity. Under this conditions, as well as in the range between Stokes law and Newton's law the terminal velocity may be calculated from equation (5.26) using a value of C_D found by trial from the Fig. 5.1.

5.7 HINDERED SETTLING

When the concentration of solids is high surrounding the particles interface with the motion of other individual particles, the influence of the neighbouring particles affect the velocity gradient surrounding each particle. This phenomena of settling particle is called Hindered settling.

When individual particles fall freely in a fluid medium without being touched by other particles is called free settling.

In case of free settling particle falls due to gravitational force in such a manner that other particles do not hinder its fall. As the particle falls, its velocity increases and will continue to increase until the accelerating force and the resisting forces are equal. When this conditions is reached, the particle velocity remains constant during the remainder of its fall unless the balance of force is upset by external force. This ultimate constant velocity is called terminal settling velocity.

When a spherical particle is allowed to fall in a stagnant fluid, it will move downwards under the influence of a gravitational force, when the bulk of the fluid with respect to particles is free, i.e., the distance between the particles is atleast 20 times the diameter then the process of settling is termed as free settling.

5.7.1 Modification of Equation for Hindered Settling

If the settling of particles is carried out at high concentration of solids to liquid are practically continuous. In this case, the terminal velocity will be entirely different. Hindered settling occurs mostly in viscous region and equation of the following type may be used to find the terminal settling velocity.

$$u_h = u_t \lambda(e) \qquad \text{...(5.37)}$$

where, u_h = Hindered settling velocity

u_t = Settling velocity

e = Porosity

$$\lambda(e) = e^2 \cdot 10^{-1.82(1-e)} \qquad \text{...(5.38)}$$

Example 5.8. The particles of spharalite (Specific gravity = 4.0) are settling under the force of gravity through slurry consisting of 25% by volume of quartz particles (specific gravity = 2.65) and water. The diameter of the spharalite particle is 0.15 mm. The volumetric ratio of sphalarite to slurry is 0.25. The temperature is 10°C. What is the Hindered settling velocity of spharalite?

Solution: From the Stoke's law we have

$$u_t = \frac{d_P^2(\rho_P - \rho)g}{18\mu}$$

Data given:

$$d_p = 0.15 \times 10^{-3} \text{ m}$$

$$\rho_P = 4000 \text{ kg/m}^3$$

$$\rho = \rho_q \times 0.25 + \rho_u \times 0.75$$

$$= 4000 \times 0.25 + 1000 \times 0.75$$

$$= 1412 \text{ kg/m}^3$$

$$\mu = 1 \times 10^{-3} \text{ kg/m.s}$$

$$g = 9.81 \text{ m/s}^2$$

$$\Rightarrow \quad u_t = \frac{(0.15\times10^{-3})^2\times(4000-1412)\times9.81}{18\times1\times10^{-3}}$$

$$\Rightarrow \quad u_t = 0.032 \text{ m/s}$$

$$e = \frac{\text{Volume of slurry} - \text{Volume of Particle}}{\text{Volume of slurry}}$$

$$= 1-\frac{\text{Volume of particle}}{\text{Volume of slurry}}$$

$$= 1 - 0.25$$

$$= 0.75$$

$$\lambda(e) = e^2 \cdot 10^{-1.82(1-e)}$$

$$= (0.75)^2 \cdot 10^{-1.82(1-0.75)}$$

$$= 0.197$$

$$u_h = u_t\lambda(e)$$

$$= 0.032 \times 0.197$$

$$= 0.0063 \text{ m/s}$$

$$= 0.63 \text{ cm/s} \quad \textbf{Ans.}$$

Example 5.9. Calculate the settling velocity for hindered settling of glass spheres at 10°C, when the suspension 1206 g of glass spheres in 1140 cm^3 of total suspension volume. The average diameter of glass sphere is 0.016 cm and density of glass is 2.54 g/cc.

Solution: From the Stokes law we have

$$u_t = \frac{d_P^2(\rho_P-\rho)g}{18\mu}$$

Data given:

$$d_P = 0.016 \times 10^{-12} \text{ m}$$

$$\rho_P = 2540 \text{ kg/m}^2$$

$$V_P = \frac{\text{Mass of glass sphere}}{\text{Density of glass sphere}}$$

$$= \frac{1206}{2.54}$$

$$= 474.8 \text{ cc}$$

$$e = 1-\frac{\text{Volume of particle}}{\text{Volume of suspension}}$$

$$= 1-\frac{474.8}{1140}$$

$$= 0.58$$

$$\rho = \rho_P \times 0.42 + \rho_w \times 0.58$$

$$= 2540 \times 0.42 + 1000 \times 0.58$$

$$= 1647 \text{ kg/m}^2$$

$$\mu = 1 \times 10^{-3} \text{ kg/m.s}$$

$$g = 9.81 \text{ m/s}^2$$

$$\Rightarrow \quad u_t = \frac{(0.016\times10^{-2})^2\times(2540-1647)\times9.81}{18\times10^{-3}}$$

$$\Rightarrow \quad u_t = 0.012 \text{ m/s}$$

$$\lambda(e) = e^2 \cdot 10^{-1.82(1-e)}$$

$$= 0.58^2 \cdot 10^{-1.82(1-0.58)}$$

$$\lambda(e) = 0.045$$

$$u_h = u_t\lambda(e)$$

$$= 0.012 \times 0.045$$

$$= 0.00054 \text{ m/s}$$

$$= 0.054 \text{ cm/s} \quad \textbf{Ans.}$$

5.8 CENTRIFUGAL SEPARATOR

A device used centrifugal force for the separation of particles from the fluid is called centrifugal separator. Centrifugal separators are of two types:

(*i*) Air cyclone separator

(*ii*) Hydro cyclone separator

5.8.1 Air Cyclone Separator

Cyclones are used primarily for the separation of solids from the fluid and utilized centrifugal force to affect the separation. Such a separation depends not only on particle size but also on particle density. Cyclone may be used to affect a separation on the basis of particle size or particle density or both.

It consists of essentially a short vertical cylinder closed by a flat or dished plate on top and a conical bottom. A typical air cyclone separator is shown in Fig. 5.4.

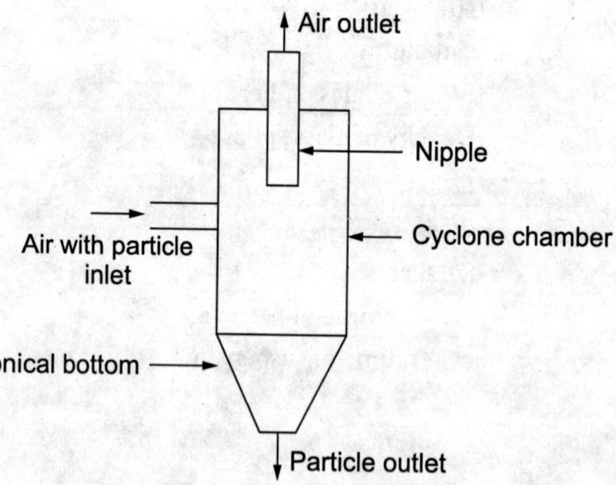

Fig. 5.4. Air cyclone separator

The air with its load of solids is introduced tangentially at the top of the cylinder portion. Centrifugal force throws the solids particle out against the wall and they drop into the hopper. The outlet for the air is usually provided with a nipple that extends inwardly into the separator to prevent the air short circuiting from the inlet to the outlet.

Such separator is widely used for the collection of wood chips, heavy and coarse dusts and all manner not too fine. They are also used for separating heavy or coarse materials from the dust.

5.8.2 Hydro Cyclone Separator

The Hydro cyclone separator is a simple piece of equipment that uses fluid pressure to generate centrifugal force and a flow pattern which can separate particles from a liquid medium. These particles must have a sufficiently different density relative to the liquid medium in order to achieve separation.

The conventional hydro cyclone consists of a cylindrical chamber connected to a conical body which leads to the bottom outlet at the apex of the cone. This discharges the underflow. The reverse flow is located by a piece which projects axially into the top of the chamber and discharges the overflow. A typical Hydro cyclone separator is shown in Fig. 5.5.

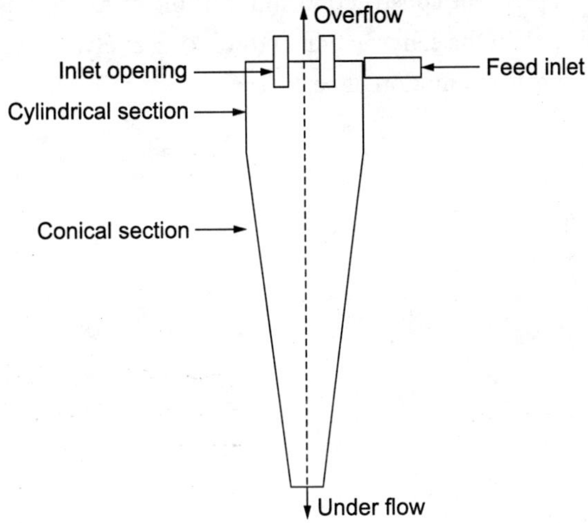

Fig. 5.5: Hydro cyclone separator

A hydro cyclone can work upside down or with an inclined axis. The existence of both downward and upward flow streams means that there is a locus of zero vertical velocity. At the same time both streams have a tangential velocity that in the outer vortex decreases with increase in radius whereas in the inner vortex decreases with decrease in radius. This means that the outer fluid is approaching a free vortex and is experiencing shear whereas the inner fluid is rotating as if it were a solid body, i.e., with constant angular velocity. The peak tangential velocity is not necessarily at the position of zero vertical velocity.

5.9 FACTORS AFFECTING THE MOTION OF PARTICLES

With an increase in temperature the particles gain kinetic energy and move faster. The actual average speed of the particles depends on their mass as well as the temperature. Heavier particles move more slowly than higher ones at the same temperature.

REVIEW QUESTIONS

5.1 Explain the mechanism of particle in motion.

5.2 Define terminal velocity.

5.3 Write briefly on drag co-efficient.

5.4 Explain the criteria for settling of particle.

5.5 What is hindered settling? Explain.

5.6 Explain the term free settling.

5.7 Explain the condition for terminal settling velocity.

5.8 What is the centrifugal separator? What is its working principle?

5.9 With a neat diagram explain the construction and working of Air cyclone separator.

5.10 With a neat diagram explain the construction and working of Hydro cyclone separator.

5.11 What are the factors affects the motion of particles?

EXERCISE

5.1 Derive the equation for one dimensional motion of particles through a fluid in gravitational field.

5.2 Derive the equation for one dimensional motion of particles through a fluid in centrifugal field.

5.3 Starting from the basic derive an equation for terminal settling velocity.

5.4 What will be the settling velocity of spherical steel ball having 0.4 mm diameter settling in an oil of specific gravity 0.80 and viscosity 0.1 poise. The density of the steel = 7850 kg/m^3.

5.5 Calculate the diameter of steel ball settling in an oil having the settling velocity 0.05 m/s. The density of the oil and steel ball are 800 kg/m^3 and 7850 kg/m^3 respectively. Assume the viscosity of oil 0.1 poise.

5.6 If the settling velocity of a spherical steel ball settling in an oil having density 800 kg/m^3 is 0.05 m/s. Calculate the density of the steel ball. The diameter of the steel ball and viscosity of oil are 0.36 mm and 0.01 kg/m.s respectively.

5.7 When the settling velocity of a spherical steel ball settling in an oil having density 800 kg/m^3 is 0.05 m/s. Calculate the viscosity of the oil in which steel ball is settling. The diameter and density of the steel ball are 0.40 mm and 7850 kg/m^3 respectively.

5.8 Estimate the terminal settling velocity for 80-100 mesh particles of limestone (ρ_p = 2800 kg/m^3) falling in water at 30°C. How much higher would be the velocity in a centrifugal separator where the acceleration is 50 g?

5.9 Particles of spharalite (specific gravity = 4.0) are settling under the force of gravity in carbon tetrachloride at 20°C (Specific gravity = 1.6). The diameter of the spharalite particle is 0.10 mm. The volume fraction of spharalite in carbon tetrachloride is 0.20. What is the settling velocity of the spharalite?

5.10 Calculate the settling velocity for the hindered settling of glass spheres in water at 20°C when the suspension contains 1296 g of glass sphere in 1150 cc of total volume. The average diameter of the sphere as determined from photomicrographs was 0.04 mm. The true density of the spheres was 2463 kg/m^3.

5.11 A steel sphere 3 mm diameter falls through kerosene covering a distance of 25 cm in 10 second. The specific gravity of steel sphere and kerosene are 7.8 and 0.8 respectively. Calculate the value of viscosity of kerosene medium employed.

5.12 Drops of oil 15 microns in diameter are to be settled from their mixture with air. The specific gravity of oil is 0.9 and the air is at 25°C and 1 atm. The settling time of 1 minute is available. How high should the chamber be to allow settling of these particles? The viscosity of air is 0.018 cp and molecular weight of air is 29.

5.13 A spherical steel ball (ρ = 7000 kg/m^3) of size 6.25 mm, takes 6.4 seconds to fall through a distance of 25 cm, in a viscous liquid. Find the kinetic viscosity of the liquid. Density of the medium is 900 kg/m^3. [VTU Exam August 1997]

❑❑❑

6 CHAPTER Sedimentation

6.1 INTRODUCTION

Sedimentation is a process during which particles suspended in water tend to settle down. This process is applicable in wide range of contexts including waste water treatment. It is important for the removal of suspended solids during the primary treatment of effluent. It is also used in the initial treatment of drinking water.

The capacity of a continuous thickener is determined by the fact that the solids initially present in the feed must be able to settle through all zones of slurry concentration from that of the initial feed to that of the underflow at a rate equal to that at which they are introduced into the thickener.

The minimum area of the thickener can be determined using the data from batch sedimentation. It is based on mathematical analysis presented by Kynch. The assumption made in this process is for a given set of operating conditions the settling rate is a function only of the solids concentration expressed as volume of solids per unit volume of slurry. If the solid handling capacity per unit area is lowest at some intermediate concentration, a zone of such concentration must start building up since the rate at which solid enters the zone will be less than the rate at which they will leave the zone.

6.2 BATCH SEDIMENTATION

In the sedimentation of concentrated suspension particles move downwards under the action of gravity and an equal volume of the fluids is displaced to the upper part. This process is best described by Batch sedimentation. A typical batch sedimentation process is shown in Fig. 6.1.

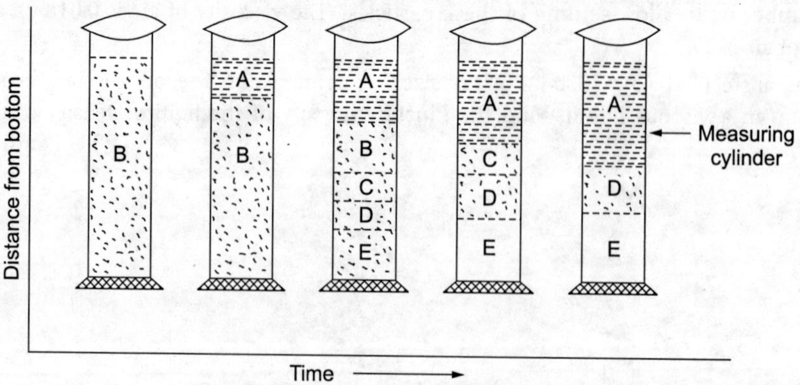

Fig. 6.1: Batch sedimentation process

When the slurry is allowed to settle under the influence of gravity, generally the following stages are formed.

A — Clear liquid

B — Suspended with original concentration

C — Transition zone through which concentration gradually increases

D — Sediment thickened slurry in compression zone

E — Sediment deposit.

The slurry is the initial concentration in the beginning of the sedimentation process. Coarse particles like sand settle down first and denoted by *E*. The interface layer C is formed next. As the particle settle down the settling of particles leaves a clear liquid phase A at the top phase. It is the slurry in original concentration.

Distance from the bottom of cylinder versus concentration may be plotted to know various stages at successive intervals. Another method of interface height versus time may be plotted to know the settling rate of particles. One such plot is shown in Fig. 6.2.

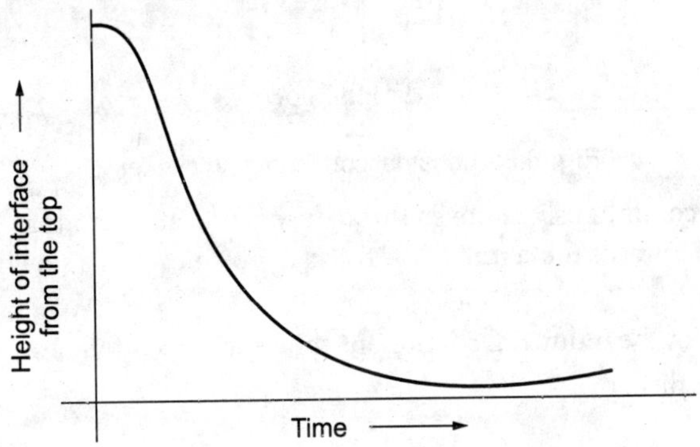

Fig. 6.2: Batch settling of slurry

6.2.1 Application of Batch Sedimentation

The capacity of a continuous thickener is determined by the fact that the solids initially present in the feed must be able to settle through all zone of slurry concentration from the initial to that of the underflow at a rate equal to that at which they are introduced into the thickener. If the area provided is not sufficient, the solids will build up through the settling zone into the clarifier zone until some solids are finally discharged in the overflow. Further more, it is not known at the start which zone will be zone of minimum capacity.

At a given set of operating condition, it is assumed that settling rate is a function only of the solid concentration expressed as volume of solids per unit volume of slurry.

6.3 INDUSTRIAL CONTINUOUS THICKENER

Dorr thickener is one of the industrial continuous thickener. It consists of a large diameter shallow tank with a flat bottom. The slurry is fed in the top at the centre into a feed well kept at a depth of 3 to 4 ft. below the surface of liquid. Feeding is done with a little disturbance as possible. A typical industrial continuous thickener is shown in Fig. 6.3.

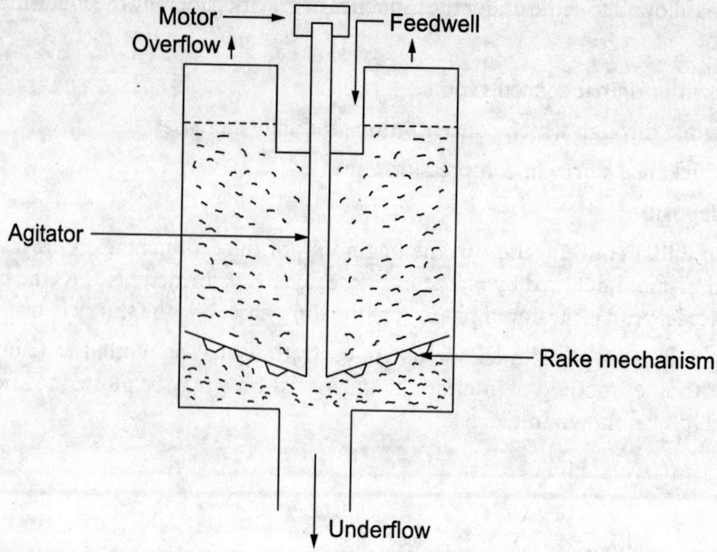

Fig. 6.3: Industrial continuous thickener

The thickened liquor is continuously removed through an outlet at the bottom. Any solid deposited in the floor of the tank are directed towards the centre of the thickener by means of a slowly rotating rake mechanism incorporating with scrapper.

The theories proposed for the industrial continuous thickener are as follows:

(*i*) Coe and Clevenger theory

(*ii*) Kynch theory

6.3.1 Coe and Clevenger Theory

The sedimentation of metallurgical slimes has been studied by Coe and Clevenger who concluded that a concentrated suspension may settle in one of the two different ways. In the first case, after an initial brief acceleration period, the interface between the clear liquid and the suspension moves downwards at a constant rate and a layer of sediment, its rate of fall decreases until the "critical settling point" is reached when a direct interface is formed between the sediment and the clear liquid. Further sedimentation results solely from a consolidation of the sediment with liquid being forced upwards around the solids which are then forming a loose bed with particle in contact with one another. Since the flow area is gradually reduced, the rate progressively diminishes. The different stages of sedimentation process is shown in Fig. 6.4.

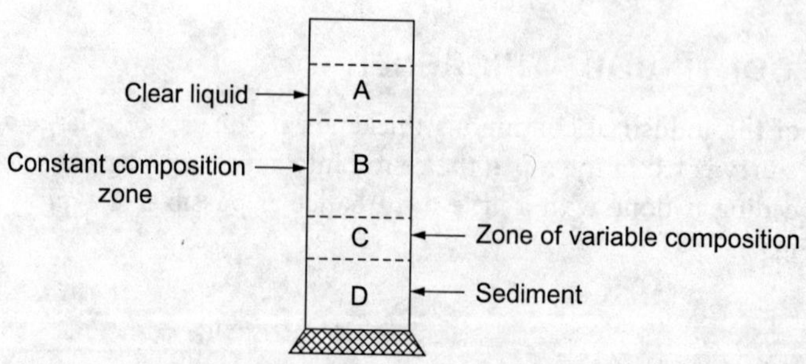

Fig. 6.4: Sedimentation of concentrated suspension with initial brief acceleration

A is clear liquid, *B* is suspension of original concentration, *C* is layer through which the concentration gradually increases, *D* is sediment thickened slurry.

The sedimentation rate remains constant until the upper interface corresponds with the top of zone *C* and it then falls, with the critical settling point is reached when both zone *B* and *C* will disappear.

In second case, rather less common mode of sedimentation is obtained when the range of particle size is very high. The sedimentation rate progressively decreases throughout the whole operation because there is no zone of constant composition and zone *C* extend from the top interface to the layer of sediment. The different stages of sedimentation for this case is shown in Fig. 6.5.

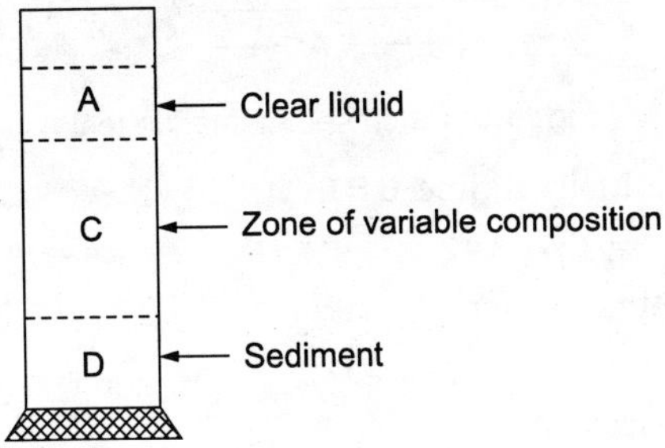

Fig. 6.5: Sedimentation of concentrated suspension with a high range of particles.

The main reason for the modification of the settling rate of particles in a concentrated suspensions are as follows:

(*i*) If a significant size range is present, the large particles are settling relative to a suspension of smaller one so that the effective density and viscosity of the fluids are increased.

(*ii*) The upwards velocity of the fluids displaced during settling is appreciable in a concentrated suspension and the apparent settling velocity is less than the actual velocity relative to the fluid.

(*iii*) The velocity gradients in the fluid close to the particles are increased as a result the change in the area and shape of the flow spaces.

(*iv*) The smaller particles tend to be dragged downwards by the motion of the large particles are therefore accelerated.

(*v*) Because the particles are closer together in a concentrated suspension, flocculation is more marked in an ionized solvent and the effective size of the small particle is increased.

If the range of particle size is not more than about 6 : 2 a concentrated suspension settle with a sharp interface and all the particle fall at the same velocity. This is in contrast with the behavior of a dilute suspension for which the rate of settling of the particle can be calculated using the more general equation.

6.3.2 Kynch Theory

Kynch has proposed a theory by mathematical analysis of the batch settling character. Here, the settling rate is determined from a single batch test when the solids handling capacity per unit area is lowest, the intermediate zone '*C*' build up as the rate at which the solids enters the zone is less than the rate at which they leave the zone. Consider a batch settling test as shown in Fig. 6.6.

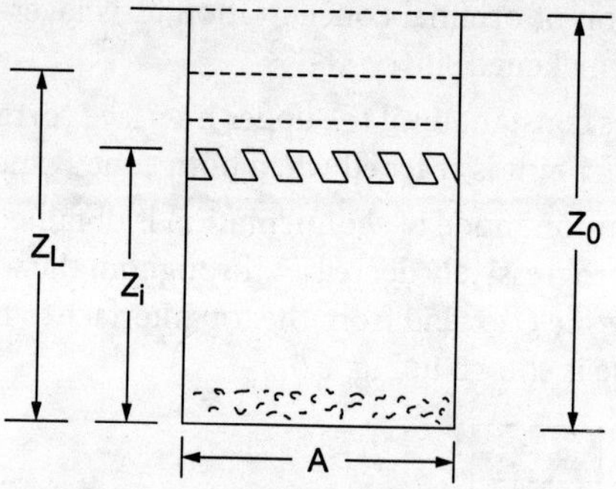

Fig. 6.6: Batch sedimentation test

The amount of solids present initially at time, $\theta = 0$

$$= C_0Z_0A \quad \text{...(6.1)}$$

where, C_0 = Initial concentration, g/cm^3

Z_0 = Initial height, cm

A = Cross-section area, cm^2

The rate of upward propagation velocity of such a zone is constant and is a function of the solid concentration. It may be expressed as

$$\bar{V} = C\frac{dv}{dc} - V \quad \text{...(6.2)}$$

where, $\bar{V}$ = Upward velocity of concentration zone at minimum settling rate with respect to vessel, cm/s

V = Settling velocity of solids in concentration zone at minimum settling rate with respect to vessel, cm/s

C = Concentration of solids (weight of solids per unit volume of pulp), g/cm^3

From the assumption that the settling rate is a function only of the solid concentration, means that

$$V = f(C) \quad \text{...(6.3)}$$

Substituting equation (6.3) in equation (6.2), we get

$$\bar{V} = Cf'(C) - f(C) \quad \text{...(6.4)}$$

Consider the test at the instant of time when the layer corresponding to the limiting settling rate has reached the interface between the clear supernant liquid and the pulp. All the solids in the initial pulp must have passed through this layer since the layer was propagated upward from the bottom of column. If the concentration of this layer is C_L and the time instant at which the layer reaches the interface is θ_L, then

$$C_0Z_0A = C_L(\bar{V}_L + V_L)\theta_L A \quad \text{...(6.5)}$$

where, V_L and $\bar{V}_L$ refer to the respective velocities for the layer having a solid concentration of C_L. Let Z_L corresponds to the height of the interface at time θ_L. Then

$$\bar{V}_L = \frac{Z_L}{\theta_L} \quad \text{...(6.6)}$$

Substituting equation (6.6) in equation (6.5), we get

$$C_0Z_0A = C_L\left(\frac{Z_L}{\theta_L} + V_L\right)\theta_L A$$

$$\Rightarrow \quad C_0Z_0 = C_L(Z_L + V_L \cdot \theta_L)$$

$$\Rightarrow \qquad C_L = \frac{C_0 Z_0}{Z_L + V_L \cdot \theta_L} \qquad ...(6.7)$$

The downwards settling velocity V_L may be expressed as

$$V_L = \frac{Z_i - Z_L}{\theta_L} \qquad ...(6.8)$$

where, Z_L = Height from which particles settling, cm

Z_i = Interface height, cm

Substituting equation (6.8) in equation (6.7), we get

$$C_L = \frac{C_0 Z_0}{Z_L + \left(\frac{Z_i - Z_L}{\theta_L}\right)\theta_L} \qquad ...(6.9)$$

$$\Rightarrow \qquad C_L = \frac{C_0 Z_0}{Z_L + Z_i - Z_L}$$

$$\Rightarrow \qquad C_L = \frac{C_0 Z_0}{Z_i}$$

$$\Rightarrow \qquad C_L \cdot Z_i = C_0 Z_0 \qquad ...(6.10)$$

The above equation states that Z_i is the height of uniform slurry concentration which control the same amounts of total solids as in the initial slurry. This concept may be clearly understood by the following Z vs θ plot as shown in Fig. 6.7.

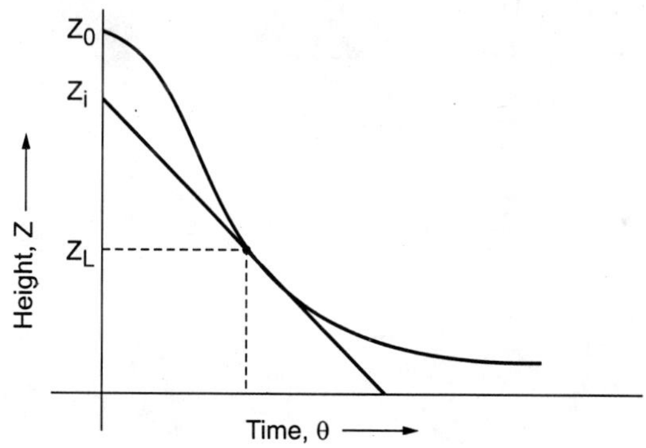

Fig. 6.7: Determination of settling velocities from batch settling curve

6.4 THICKENER DESIGN

Consider an industrial continuous thickener as shown in Fig. 6.8.

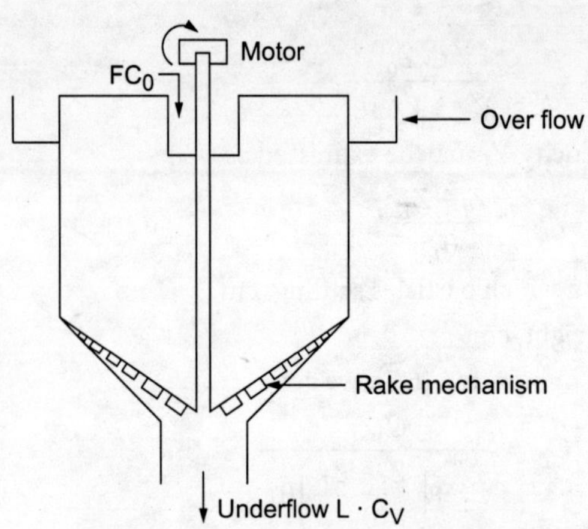

Fig. 6.8: Industrial continuous thickener

The equation $C_L \cdot Z_i = C_0 Z_0$ can be used to find the settling velocity at different time by arbitrarily fixing the time and drawing the tangent to the curve at any point and finding the slope and intercept.

Overall material balance for the thickener

$$FC_0 = LC_V \quad \text{...(6.11)}$$

$$\Rightarrow \quad L = \frac{FC_0}{C_V} \quad \text{...(6.12)}$$

Material balance for liquid

$$F(1 - C_0) = L(1 - C_V) + V \quad \text{...(6.13)}$$

Substituting equation (6.12) in equation (6.13), we get

$$F(1 - C_0) = \frac{FC_0}{C_V}(1 - C_V) + V \quad \text{...(6.14)}$$

$$\Rightarrow \quad F(1 - C_0) = \frac{FC_0}{C_V} - FC_0 + V$$

$$\Rightarrow \quad F - FC_0 = \frac{FC_0}{C_V} - FC_0 + V$$

$$\Rightarrow \quad F = \frac{FC_0}{C_V} + V$$

$$\Rightarrow \quad V = F - \frac{FC_0}{C_V}$$

$$\Rightarrow \quad V = FC_0\left[\frac{1}{C_0} - \frac{1}{C_V}\right] \quad \text{...(6.15)}$$

Let the cross-sectional area of the thickener be 'A'. Dividing equation (6.15) by A, we get

$$\frac{V}{A} = \frac{FC_0}{A}\left[\frac{1}{C_0} - \frac{1}{C_V}\right] \quad \text{...(6.16)}$$

The term V/A is the upward liquid volume in the clarification zone. Hence, the term V/A may be replaced by V_L.

$$\therefore \quad V_L = \frac{FC_0}{A}\left[\frac{1}{C_0} - \frac{1}{C_V}\right] \quad \text{...(6.17)}$$

But, $$FC_0 = L_L C_L \quad \text{...(6.18)}$$

Substituting equation (6.18) in equation (6.17), we get

$$V_L = \frac{L_L C_L}{A}\left[\frac{1}{C_0} - \frac{1}{C_V}\right]$$

$$\Rightarrow \quad \frac{L_L C_L}{A} = \frac{V_L}{[1/C_0 - 1/C_V]} \quad \text{...(6.19)}$$

where, $C_L \cdot L_L$ = Solid handling capacity, kg/s

A = Cross-sectional area, m^2

V_L = Settling velocity, m/s

C_0 = Initial concentration, kg/m^3

C_V = Underflow concentration, kg/m^3

In equation (6.19), A represent the minimum thickener area. Replacing A by S in equation (6.19), we get

$$\frac{L_L C_L}{S} = \frac{V_L}{1/C_0 - 1/C_V} \quad \text{...(6.20)}$$

This is the design equation of industrial continuous thickener. This equation can be used to calculate the minimum thickener area 'S'.

Example 6.1. A single batch settling test was made on a liquid slurry. The interface between clear liquid and suspension solids were observed as a function of time and the results are tabulated below. The test was made using 236 g of limestone per litre of slurry.

Sl. No.	Interface Height (cm)	Time (hr)
1	36.0	0.00
2	32.4	0.25
3	28.6	0.50
4	21.0	1.00
5	14.7	1.75
6	12.3	3.00
7	11.6	4.75
8	9.8	12.00
9	8.8	20.00

Prepare a curve showing the relationship between settling rate and the solid concentration. Using this plot a curve of settling velocity vs solid concentration. Find the thickener area if the slurry is fed at a rate of 50,000 kg solids/hr to produce a thickened sludge of 550 g limestone per litre. **[VTU Exam, Nov. 2007]**

Solution: Data given: $L_L \cdot C_L$ = 50,000 kg/hr

C_V = 550 g/lit

Table 6.1: Batch settling test data

Sl. No.	Interface Height (cm)	Time (hr)
1	36.0	0.00
2	32.4	0.25
3	28.6	0.50
4	21.0	1.00
5	14.7	1.75
6	12.3	3.00
7	11.6	4.75
8	9.8	12.00
9	8.8	20.00

Using the Batch settling test data given in Table 6.1 plot the graph interface height vs time as shown in Fig. 6.9.

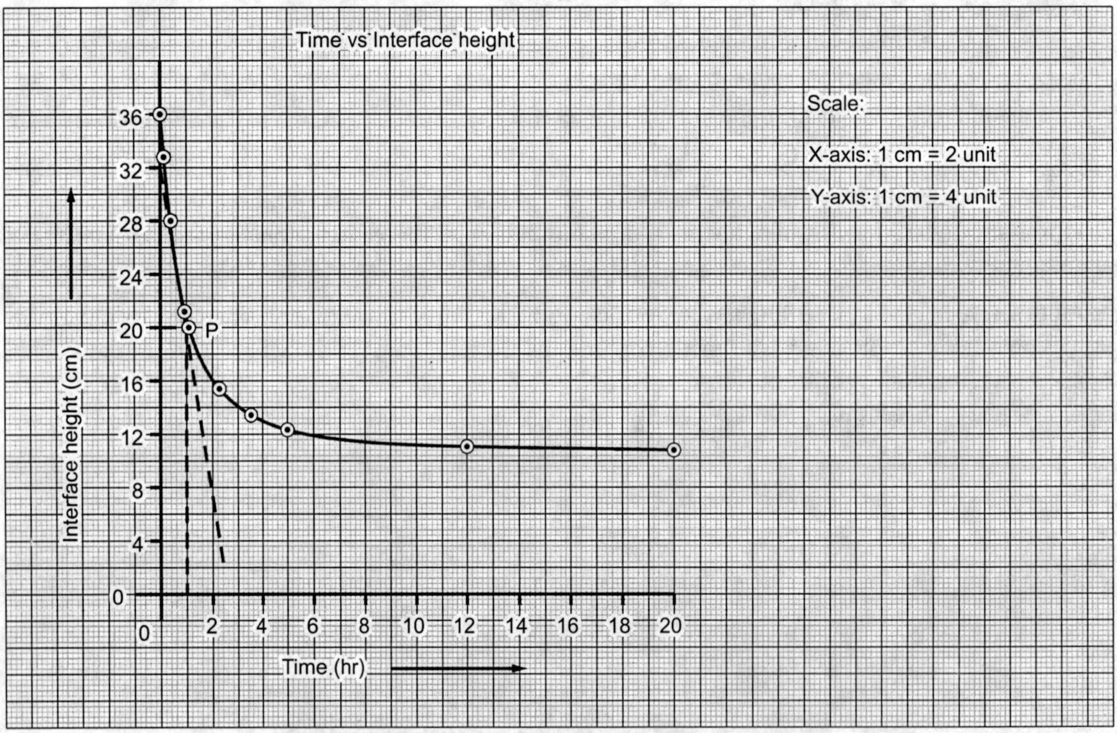

Fig. 6.9: Variation of interface height with respect to time.

Select a arbitrary point '*p*' on the curve shown in Fig. 6.9. From this point draw a horizontal line towards the *y*-axis. This will meet the *y*-axis at a particular point. Record this value as Z_L.

From the point '*p*' draw a vertical line downward towards the *x*-axis. This will meet the *x*-axis at a particular point. Record this value as θ_L.

Again from the point '*p*' draw a tangent and extend the line of tangent till it meet the *y*-axis at a particular point. Record this value as Z_i.

Repeat the same procedure to obtain a set of values as shown in Table 6.2.

Table 6.2: Arbitrarily chosen values of Z_L, θ_L and Z_i from the graph

Sl. No.	Z_L (cm)	θ_L (hr)	Z_i (cm)	$V_L = (Z_i - Z_L)/\theta_L$ (cm/hr)	$C_L = (C_0 Z_0)/Z_i$ (g/l)
1	20	1	320	12.0	265.5
2	14	2	19.5	2.75	435.7
3	12	3	16.0	1.34	531.0
4	11	4	14.0	0.75	607.0
5	10	8	12.0	0.25	708.0

Using the arbitrarily obtained values of Z_L, θ_L and Z_i, perform the calculation and get the corresponding values of V_L and C_L. Then, plot the graph C_L vs V_L as shown in Fig. 6.10.

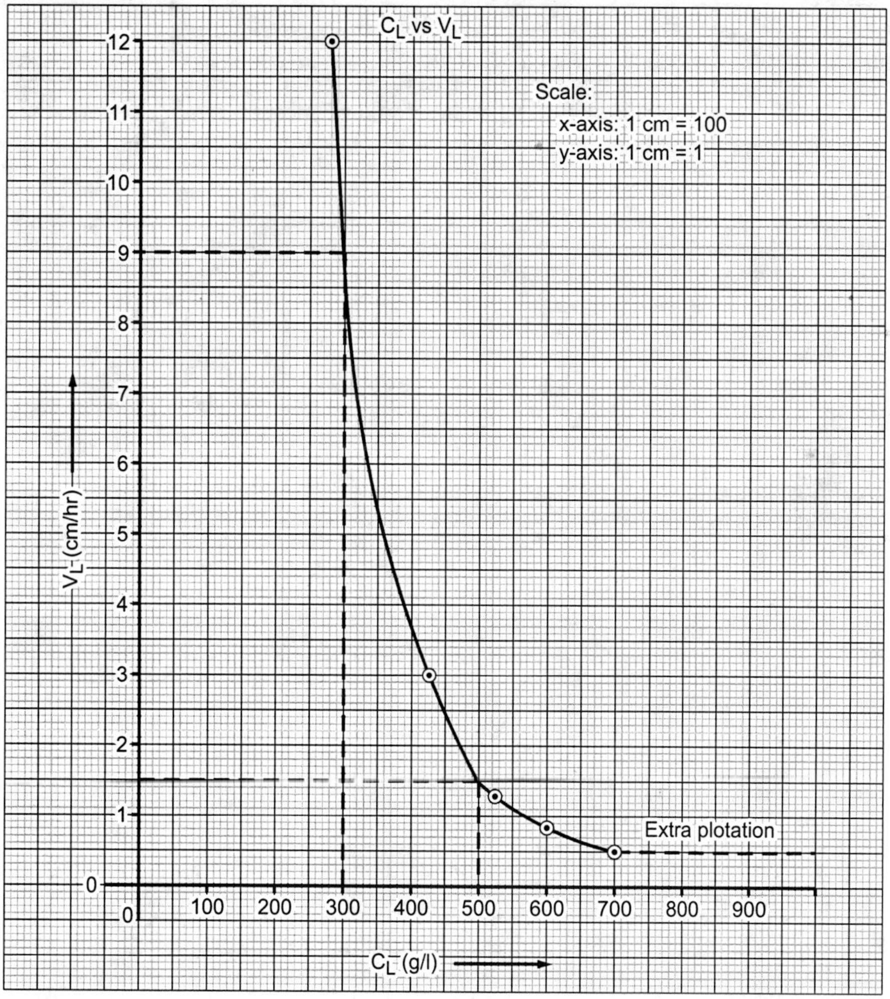

Fig. 6.10: Variation of concentration with respect to settling velocity.

The bottom part of the graph in Fig. 6.10 is very short compared to top part. Therefore, to determine the minimum thickener area, it is required to extra plot the bottom part of the graph. Such an extra plotation of the graph is shown by dotted line Fig. 6.10.

Select arbitrarily the value of V_L from the graph shown in Fig. 6.10 and obtain the corresponding value of C_L. Repeat the same procedure to obtain a set of values as shown in Table 6.3.

Table 6.3: Arbitrarily chosen value of V_L, C_L from the graph

Sl. No.	V_L(cm/hr)	C_L(g/l)	C_V(g/l)	$1/C_L$(l/g)	$1/C_V$(l/g)	$(1/C_L - 1/C_V)$ (l/g)	$\frac{L_L \cdot C_L}{S} = \frac{V_L}{1/C_0 - 1/C_V}$ (cm/hr)(g/l)
1	9	300	550	0.0033	0.0018	0.0015	6000
2	6	350	550	0.0025	0.0018	0.0011	5455
3	4.3	400	550	0.0025	0.0018	0.0007	6143
4	2.5	450	550	0.0022	0.0018	0.0004	6250
5	1.5	500	550	0.0020	0.0018	0.0002	7500

Using the arbitrarily obtained values of V_L, C_L and underflow concentration C_V, perform the calculation and get the corresponding value of $L_L \cdot C_L/S$. Then, plot the graph $V_L V_S (L_L \cdot C_L/S)$ as shown in Fig. 6.11.

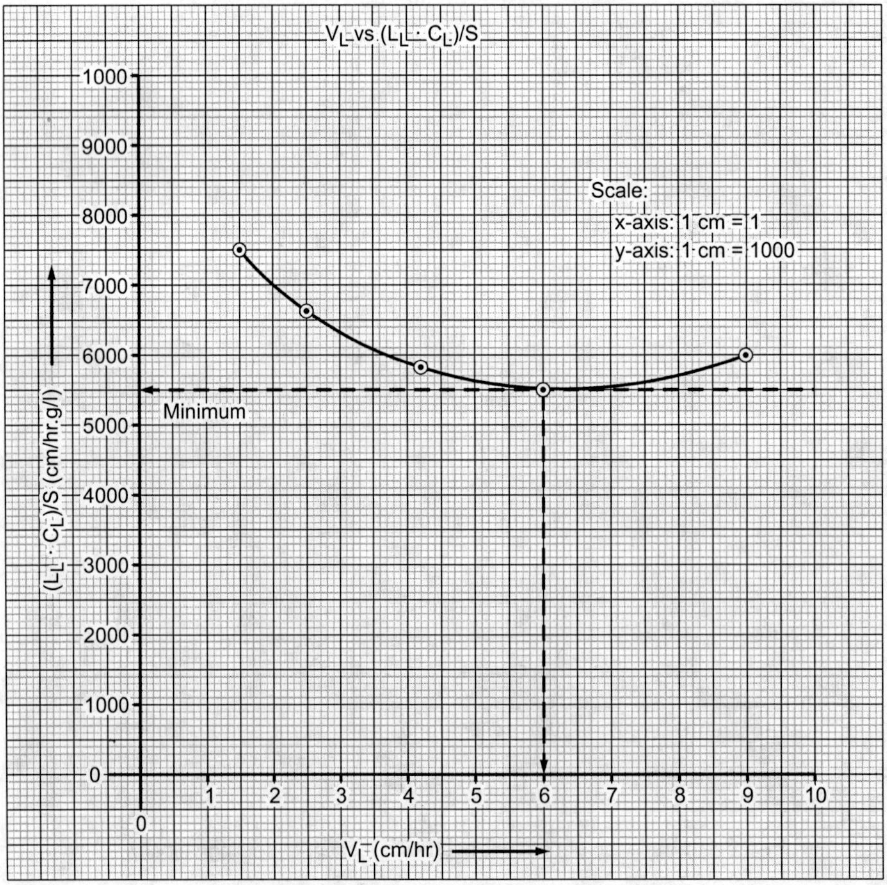

Fig. 6.11 Minimum value of $(L_L \cdot C_L/S)$

From the equation of minimum thickener area, we have

$$\frac{L_L \cdot C_L}{S} = \frac{V_L}{(1/C_L - 1/C_V)}$$

$$\Rightarrow \quad \frac{L_L \cdot C_L}{S} = 5455 \text{ cm/hr} \cdot \text{g/l}$$

$$\Rightarrow \quad S = \frac{50{,}000 \times 1000 \times 1000}{5455 \times 10000}$$

$$\Rightarrow \quad S = 916 \text{ m}^2$$

Minimum thickener area is 916 m^2. **Ans.**

Example 6.2. 100 tonnes/hr of suspension is to be concentrated from 186 g/l to 1200 g/l in Dorr thickener. Determine the minimum area of thickener using the following batch sedimentation test results. Specific gravity of solid = 4.44.

Sl. No.	Pulp height (cm)	Time (hr)
1	91	0.00
2	62	0.10
3	43	0.25
4	28	0.50
5	21	0.75
6	15	1.00
7	8	2.00
8	3	4.00

[VTU Exam, Non 2006]

Solution: Data given: $L_L \cdot C_L = 1{,}00{,}000$ kg/hr

$C_V = 1200$ g/l

Specific gravity of solids = 4.44.

Table 6.4 Batch settling test data

Sl. No.	Pulp height (cm)	Time (hr)
1	91	0.00
2	62	0.10
3	43	0.25
4	28	0.50
5	21	0.75
6	15	1.00
7	8	2.00
8	3	4.00

Using the Batch settling test data given in Table 6.4 plot the graph Interface height vs time as shown in Fig. 6.12.

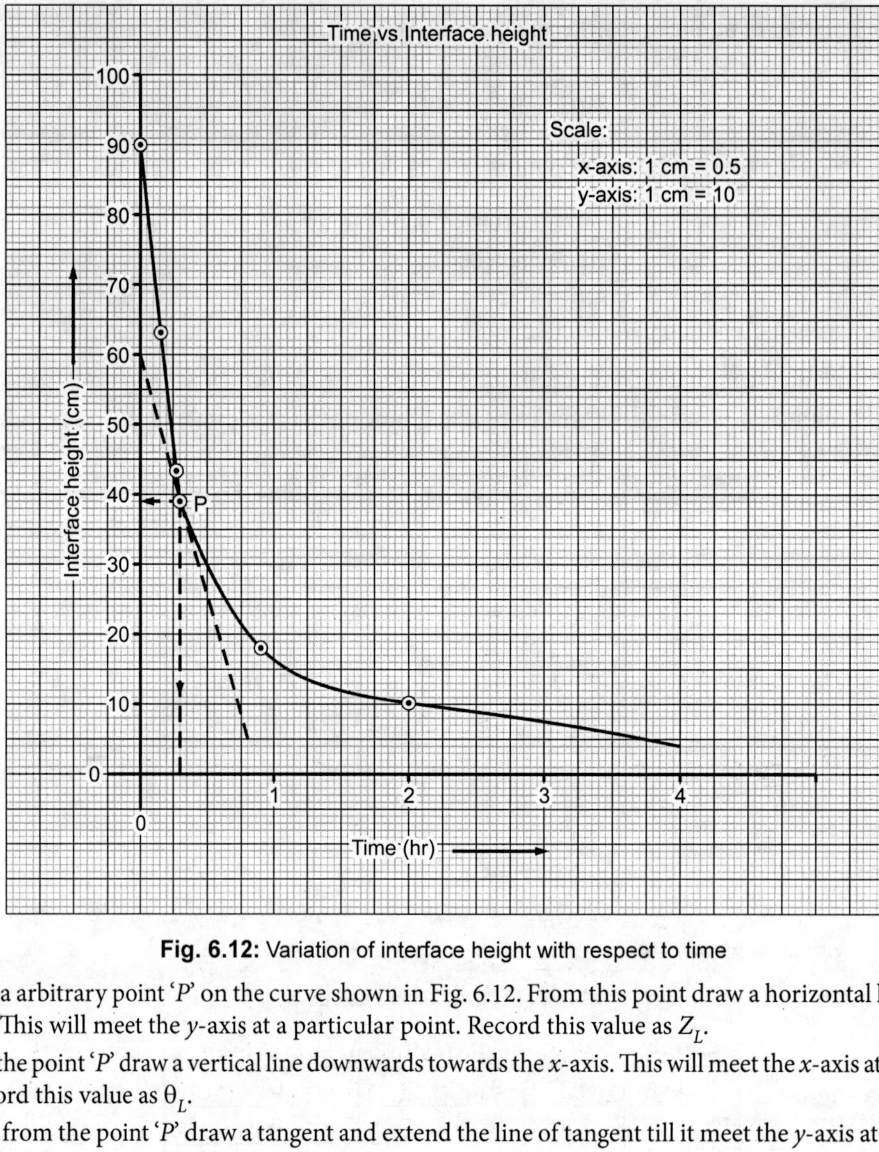

Fig. 6.12: Variation of interface height with respect to time

Select a arbitrary point '*P*' on the curve shown in Fig. 6.12. From this point draw a horizontal line towards the *y*-axis. This will meet the *y*-axis at a particular point. Record this value as Z_L.

From the point '*P*' draw a vertical line downwards towards the *x*-axis. This will meet the *x*-axis at a particular point. Record this value as θ_L.

Again from the point '*P*' draw a tangent and extend the line of tangent till it meet the *y*-axis at a particular point. Record this value as Z_i.

Repeat the same procedure to obtain a set of values as shown in Table 6.5.

Table 6.5: Arbitrarily chosen values of Z_L, θ_L and Z_i from the graph.

Sl. No.	Z_L (cm)	θ_L (hr)	Z_i (cm)	$V_L = (Z_i - Z_L)/\theta_L$ (cm/hr)	$C_L = (C_0 Z_0)/Z_i$ (g/l)
1	38	0.30	58	66.67	291.8
2	30	0.45	49	42.22	345.4
3	12	1.25	23	8.80	735.0
4	03	1.75	18	5.14	940.3
5	04	3.15	11	2.22	1538.7

Using the arbitrarily obtained values of Z_L, θ_L and Z_i. Perform the calculation and get the corresponding values of V_L and C_L. Then, plot the graph C_L vs V_L as shown in Fig. 6.13.

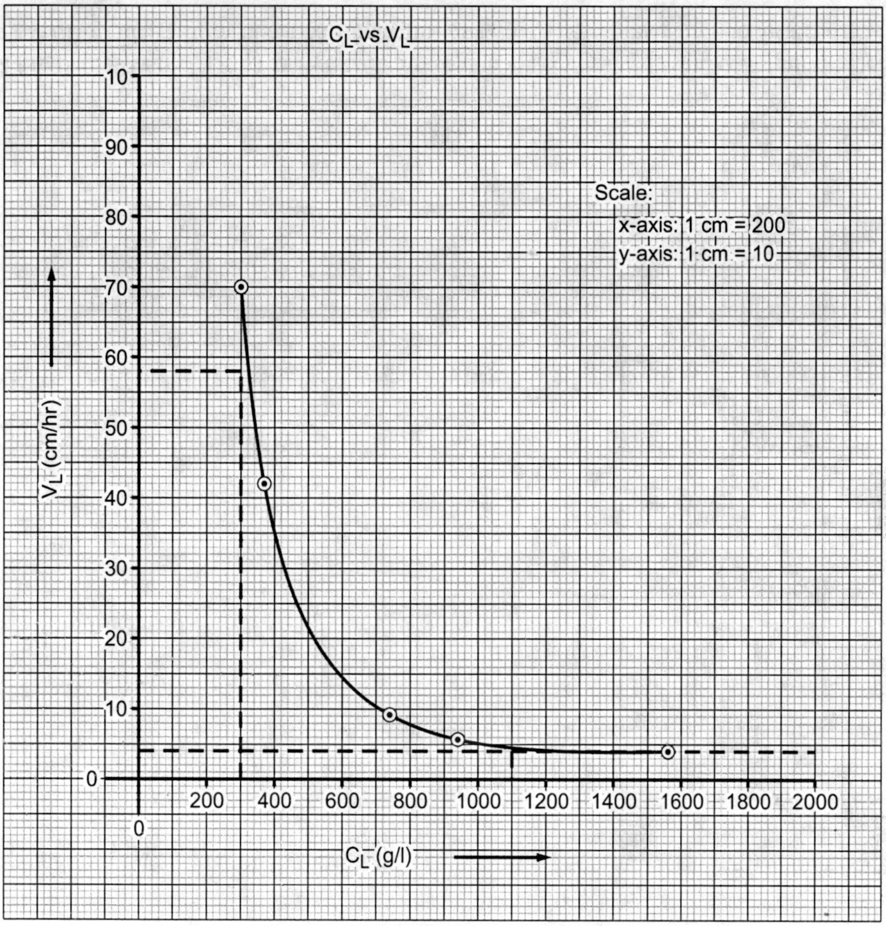

Fig. 6.13: Variation of concentration with respect to settling velocity

The bottom part of the graph in Fig. 6.13 is very short compared to top part. Therefore, to determine the minimum thickener area, it is required to extra plot the bottom part of the graph. Such an extra plotation of the graph is shown in Fig. 6.13.

Select arbitrarily the value of V_L from the graph shown in Fig. 6.13 and obtain the corresponding value of C_L. Repeat the same procedure to obtain a set of values as shown in Table 6.6.

Table 6.6: Arbitrarily chosen values of V_L, C_L from the graph

Sl. No.	V_L(cm/hr)	C_L (g/l)	C_V(g/l)	$1/C_L$ (l/g)	$1/C_V$ (l/g)	$(1/C_L - 1/C_V)$ (l/g)	$\frac{L_L \cdot C_L}{S} = \frac{V_L}{(1/C_L \cdot 1/C_V)}$ (cm/hr)(g/l)
1	58.0	300	1200	0.0033	0.00083	0.00247	23,482
2	33.5	400	1200	0.0025	0.00083	0.00167	20,060
3	11.0	700	1200	0.0014	0.00083	0.00057	19,298

4	5.5	900	1200	0.0011	0.00083	0.00027	20,370
5	3.5	1100	1200	0.0009	0.00083	0.00007	50,000

Using the arbitrarily obtained values of V_L, C_L and underflow concentration C_V, perform the calculation and get the corresponding values of $L_L \cdot C_L/S$. Then, plot the graph V_L vs $(L_L \cdot C_L/S)$ as shown in Fig. 6.14.

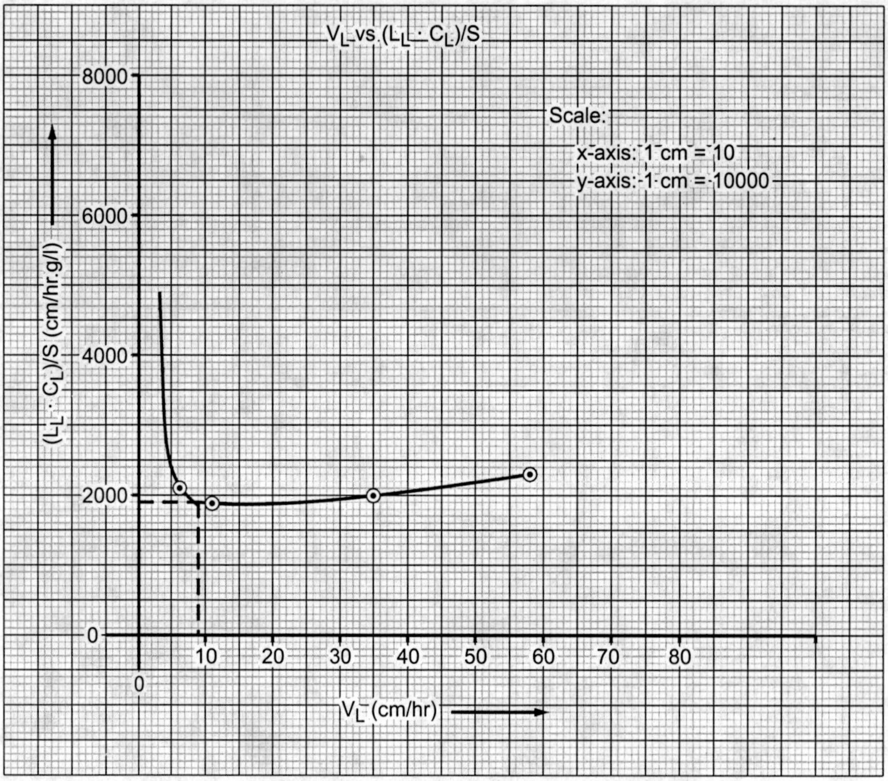

Fig. 6.14: Minimum value of $L_L \cdot C_L/S$

From the equation of minimum thickener area, we have

$$\frac{L_L \cdot C_L}{S} = \frac{V_L}{1/C_L - 1/C_V}$$

$$\Rightarrow \quad \frac{L_L \cdot C_L}{S} = 18{,}000 \text{ (cm/hr) (g/l)}$$

$$\Rightarrow \quad S = \frac{100{,}000 \times 1000 \times 1000}{18000 \times 10000}$$

$$\Rightarrow \quad S = 556 \text{ m}^2$$

The minimum thickener area is 556 m^2. **Ans.**

Example 6.3. Determine the minimum cross-sectional area of a continuous thickener required to handle 150 ton/day of dry solids from initial concentration of 5% to a given underflow concentration of 50% by conducting a batch sedimentation test. The test datas are given below:

Sl. No.	Height (cm)	Time (s)
1	33.5	0
2	32.5	20
3	31.5	51
4	30.5	65
5	29.5	81
6	28.5	92
7	27.5	117
8	26.5	135
9	25.5	150
10	24.5	170
11	23.0	187
12	22.5	200
13	21.0	217
14	20.5	230
15	18.5	250
16	17.5	270
17	17.0	277
18	16.0	300
19	15.0	307
20	14.0	315
21	13.5	330
22	12.5	340
23	11.5	358
24	11.0	373
25	10.0	394
26	9.0	407
27	8.5	438
28	8.0	480
29	7.5	525
30	7.0	555
31	6.5	604
32	6.0	735
33	5.5	828
34	5.0	1010
35	4.5	1415
36	4.0	2238

Initial height of the slurry, $Z_0 = 33.5$ cm

Density of $CaCO_3 = 2560$ kg/m^3

Solution: Data given: $L_L \cdot C_L = 150$ ton/day

Initial concentration, $C_0 = \dfrac{\text{Weight of solid}}{\text{Volume of solids + Volume of water}}$

$$= \frac{50\times10^{-3}}{\left(\dfrac{50\times10^{-3}}{2560}\right)+(950\times10^{-6})}$$

$$= 51.59\ \text{kg/m}^3$$

Underflow concentration, $C_V = \dfrac{\text{Weight of solid}}{\text{Volume of solid + Volume of water}}$

$$= \frac{500\times10^{-3}}{\left(\dfrac{500\times10^{-3}}{2560}\right)+500\times10^{-6}}$$

$$= 719.11\ \text{kg/m}^3$$

Table 6.7: Batch settling test data

Sl. No.	Height (cm)	Time (s)
1	33.5	0
2	32.5	20
3	31.5	51
4	30.5	65
5	29.5	81
6	28.5	92
7	27.5	117
8	26.5	135
9	25.5	150
10	24.5	170
11	23.0	187
12	22.5	200
13	21.0	217
14	20.5	230
15	18.5	250
16	17.5	270
17	17.0	277
18	16.0	300
19	15.0	307
20	14.0	315
21	13.5	330
22	12.5	340
23	11.5	358
24	11.0	373

Sl. No.	Height (cm)	Time (s)
25	10.0	394
26	9.0	407
27	8.5	438
28	8.0	480
29	7.5	525
30	7.0	555
31	6.5	664
32	6.0	735
33	5.5	828
34	5.0	1010
35	4.5	1415
36	4.0	2238

Using the batch settling test data given in Table 6.7 plot the graph interface height vs time as shown in Fig. 6.15.

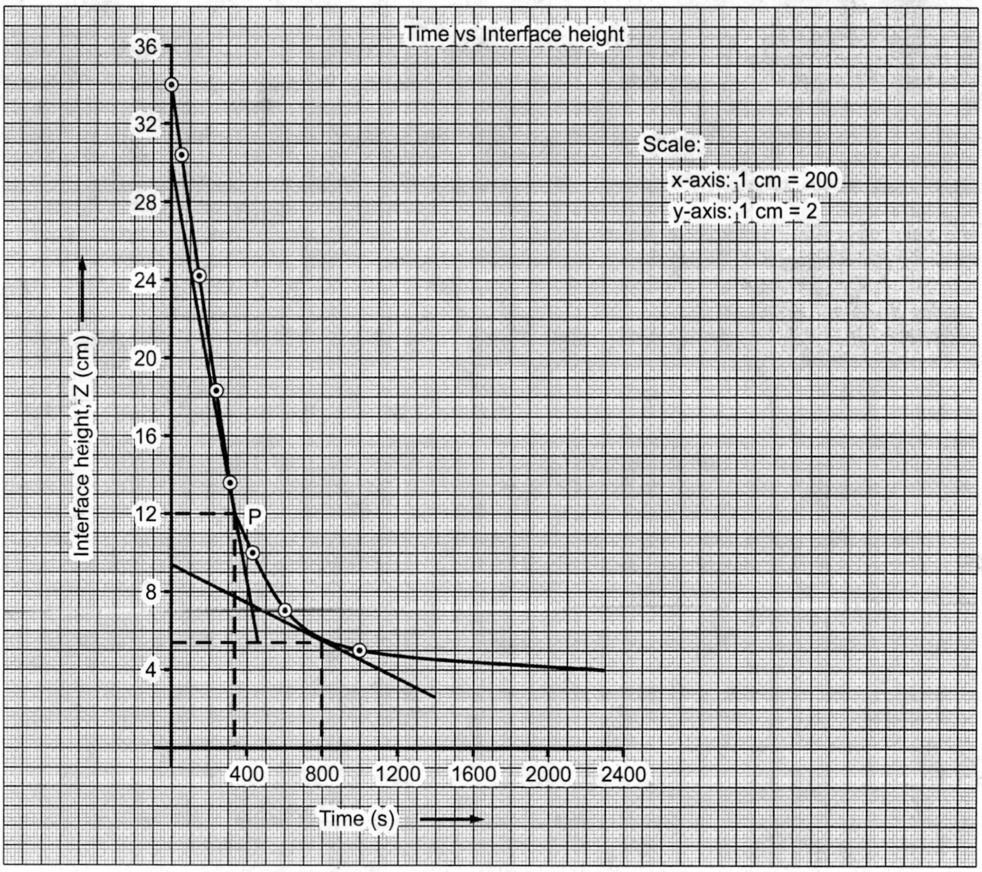

Fig. 6.15: Variation of interface height with respect to time.

Select a arbitrary point '*P*' on the curve shown in Fig. 6.15. From this point draw a horizontal line towards the *y*-axis. This will meet the *y*-axis at a particular point. Record this value as Z_L.

From the point '*P*' draw a vertical line downwards towards the *x*-axis. This will meet the *x*-axis at a particular point. Record this value as θ_L.

Again from the point '*P*' draw a tangent and extend the line of tangent till it meet the *y*-axis at a particular point. Record this value as Z_i. Repeat the same procedure to obtain a set of values as shown in Table 6.8.

Table 6.8: Arbitrarily chosen value of V_L, θ_L and Z_i from the graph

Sl. No.	Z_L (cm)	θ_L (s)	Z_i (cm)	$V_L = (Z_i - Z_L)/\theta_L$ (m/s)	$C_L = (C_0Z_0)/Z_i$ (kg/m³)
1	12.2	340	29.0	4.94×10^{-4}	59.59
2	9.8	400	22.8	3.25×10^{-4}	77.57
3	7.6	515	14.6	1.36×10^{-4}	113.37
4	6.6	590	10.8	0.71×10^{-4}	160.02
5	5.6	800	9.2	0.45×10^{-4}	187.86

Using the arbitrarily obtained values of Z_L, θ_L and Z_i, perform the calculation and get the corresponding values of V_L and C_L. Then, plot the graph C_L vs V_L as shown in Fig. 6.16.

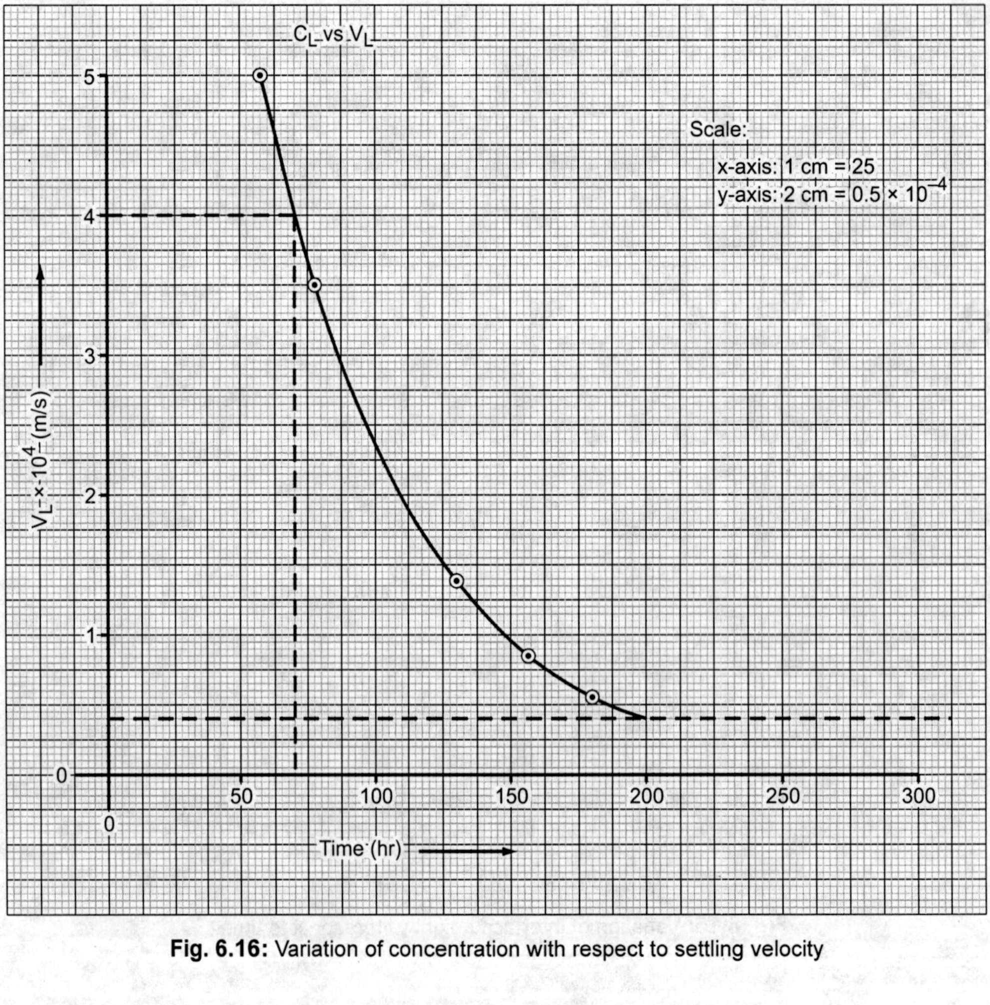

Fig. 6.16: Variation of concentration with respect to settling velocity

The bottom part of the graph in Fig. 6.16 is very short compared to top part. Therefore, to determine the minimum thickener area it is required to extra plot the bottom part of the graph. Such an extraplotation of the graph is shown by dotted line in Fig. 6.16.

Select arbitrarily the value of V_L from the graph shown in Fig. 6.16 and obtain the corresponding value of C_L. Repeat the same procedure to obtain a set of values as shown in Table 6.9.

Table 6.9: Arbitrarily chosen value of V_L, C_L from the graph

Sl. No.	V_L (m/s)	C_L (kg/m³)	C_V (kg/m³)	$1/C_L$ (m³/kg)	$1/C_V$ (m³/kg)	$(1/C_L - 1/C_V)$ (m³/kg)	$\frac{L_L \cdot C_L}{S} = \frac{V_L}{(1/C_L - 1/C_V)}$ (m/s)(kg/m³)
1	4×10^{-4}	67.5	719.11	0.0148	0.0014	0.0134	0.0298
2	3×10^{-4}	78.8	719.11	0.0127	0.0014	0.0113	0.0265
3	2×10^{-4}	100.0	719.11	0.0100	0.0014	0.0086	0.0233
4	1.5×10^{-4}	131.3	719.11	0.0076	0.0014	0.0062	0.0161
5	0.3×10^{-4}	320.0	719.11	0.0031	0.0014	0.0017	0.0176

Using the arbitrarily obtained values of V_L, C_L and under flow concentration C_V, perform the calculation and get the corresponding value of $L_L \cdot C_L/S$. Then, plot the graph V_L vs $(L_L \cdot C_L/S)$ as shown in Fig. 6.17.

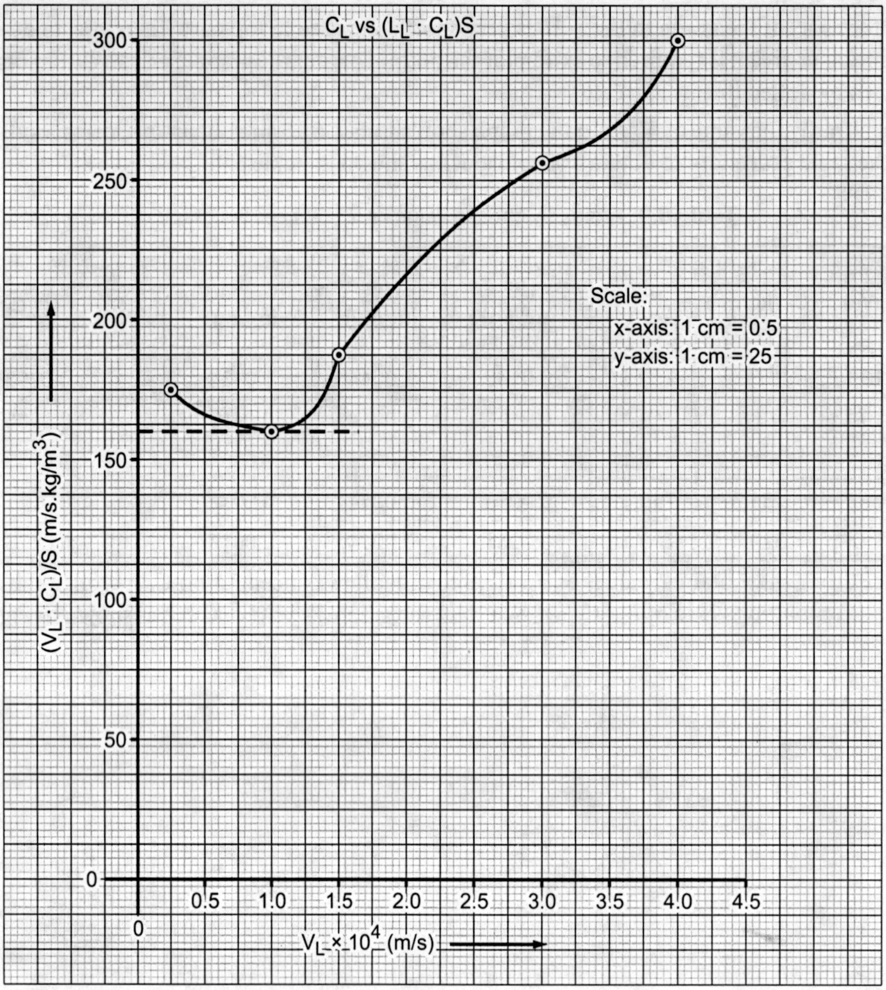

Fig. 6.17: Minimum value of $L_L \cdot C_L/S$

From the equation of minimum thickener area, we have

$$\frac{L_L \cdot C_L}{S} = \frac{V_L}{(1/C_L \cdot 1/C_V)}$$

$$\frac{L_L \cdot C_L}{S} = 161 \times 10^{-4}$$

$$S = \frac{150 \times 1000}{(24 \times 60 \times 60) \times 161 \times 10^{-4}}$$

$$S = 107.83 \text{ m}^2$$

The minimum thickener area is 107.83 m^2. **Ans.**

Example 6.4. A 3% by weight $CaCO_3$ slurry was subjected to a batch sedimentation test. The density of solids was 2.63 g/cc and that of the liquid was 1 g/cc. The test results are indicated below.

Sl. No.	*Z* (cm)	θ (min.)
1	176	0
2	100	20
3	74	40
4	57	60
5	42	80
6	34	100
7	26	120
8	18	140

where, Z = Height of settling interface, cm

θ = Settling time, min.

Determine the maximum cross-sectional area of a continuous thickener that is to handle, 4,00,000 kg of solids per day, if the initial solid concentration in the feed is 3% by weight and the final concentration is to 40% by weight of solids.

Solution: Data given:

$$C_0 = \frac{\text{Weight of solids}}{\text{Volume of solid} + \text{Volume of water}}$$

$$= \frac{30 \times 10^{-3}}{\left(\dfrac{30 \times 10^{-3}}{2630}\right) + (970 \times 10^{-6})}$$

$$= 30.57 \text{ kg/m}^3$$

$$C_V = \frac{\text{Weight of solids}}{\text{Volume of solid} + \text{Volume of water}}$$

$$= \frac{400 \times 10^{-3}}{\left(\dfrac{400 \times 10^{-3}}{2630}\right) + (600 \times 10^{-6})}$$

$$= 531.85 \text{ kg/m}^3$$

Table 6.10: Batch settling test data

Sl. No.	*Z* (cm)	θ (min.)
1	176	0
2	100	20
3	74	40
4	57	60
5	42	80
6	34	100
7	26	120
8	18	140

Using the batch settling test data given in Table 6.10. Plot the graph interface height vs time as shown in Fig. 6.18.

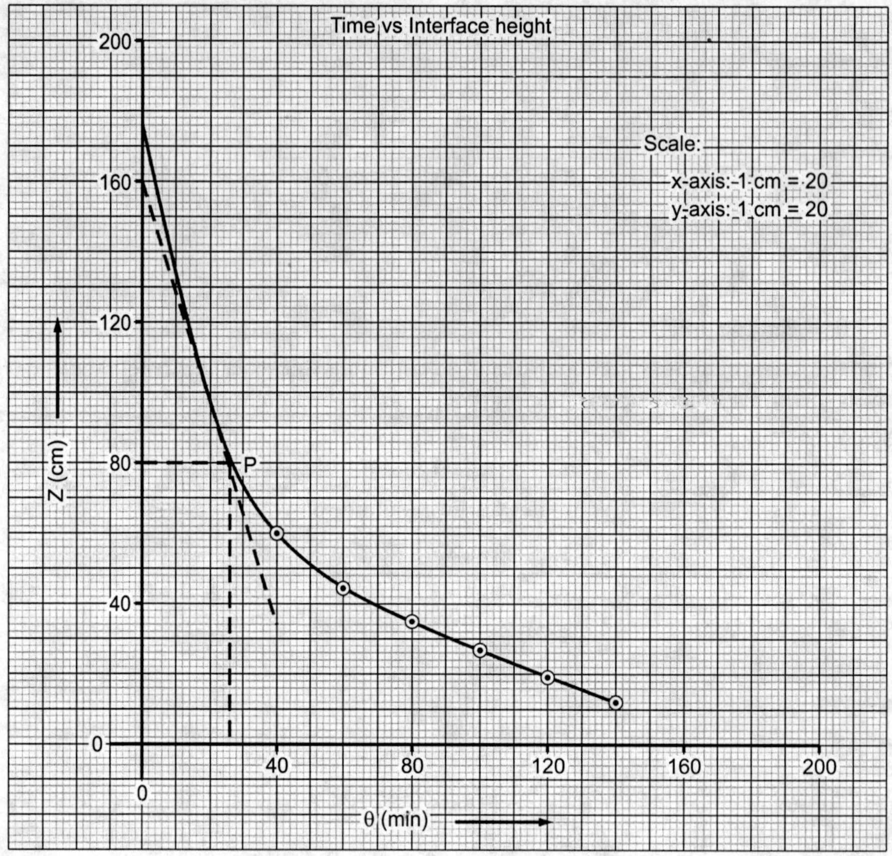

Fig. 6.18: Variation of interface height with respect to time

Select a arbitrary point '*P*' on the curve shown in Fig. 6.18. From this point draw a horizontal line towards the *y*-axis. This will meet the *y*-axis at a particular point. Record this value as Z_L.

From the point '*P*' draw a vertical line downwards toward the *x*-axis. This will meet the *x*-axis at a particular point. Record this value as θ_L.

Again from the point '*P*' draw a tangent and extend the line of tangent till it meet the *y*-axis at a particular point. Record this value as Z_i. Repeat the same procedure to obtain a set of values as shown in Table 6.11.

Table 6.11: Arbitrarily chosen value of V_L, θ_L and Z_i from the graph

Sl. No.	Z_L (cm)	θ_L (min.)	Z_i (cm)	$V_L = (Z_i - Z_L)/\theta_L$ (m/s)	$C_L = (C_0 Z_0)/Z_i$ (kg/m^3)
1	80	26	160	5.13×10^{-2}	33.63
2	60	40	92	1.33×10^{-2}	58.48
3	40	77	80	0.87×10^{-2}	67.25
4	30	88	66	0.68×10^{-2}	81.52
5	20	120	56	0.50×10^{-2}	96.08

Using the arbitrarily obtained values of Z_L, θ_L and Z_i, perform the calculation and get the corresponding values of V_L and C_L. Then, plot the graph C_L vs V_L as shown in Fig. 6.19.

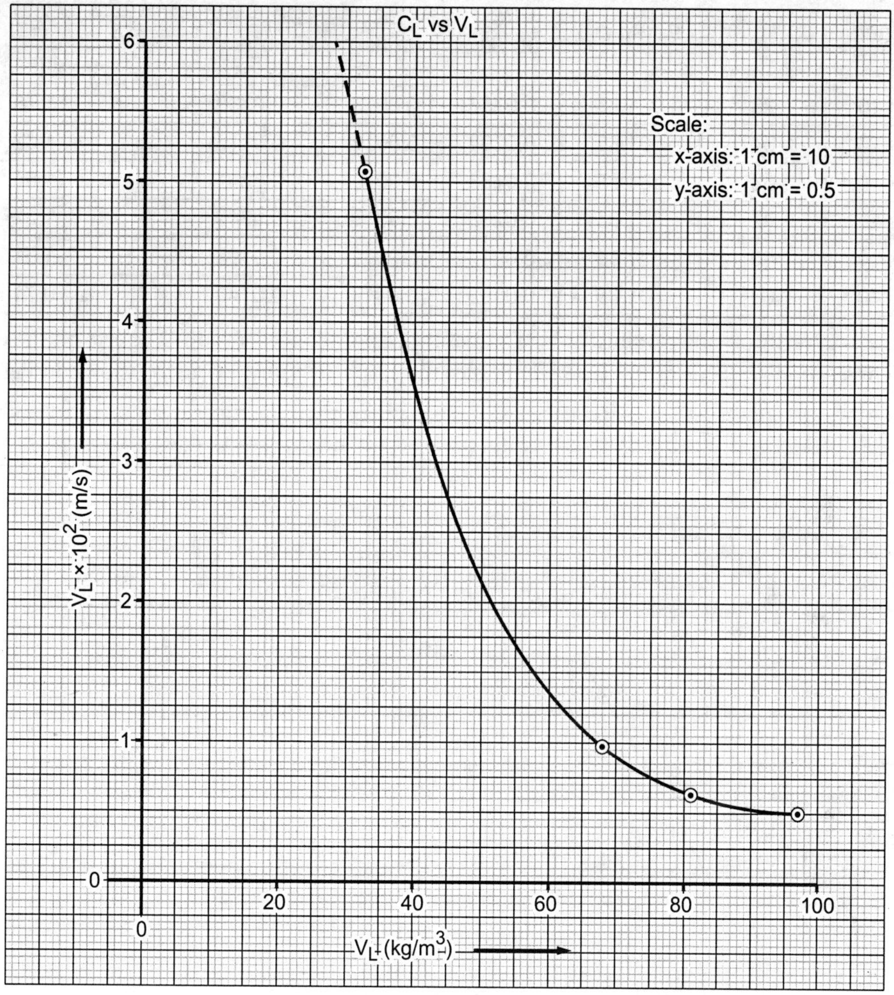

Fig. 6.19: Variation of concentration with respect to settling velocity

The top part of the graph in Fig. 6.19 is very short compared to bottom part. Therefore, to determine the maximum thickener area it is required to extra plot the top part of the graph. Such an extra plotation of the graph is shown by dotted line in Fig. 6.19.

Select arbitrarily the value of V_L from the graph shown in Fig. 6.19 and obtain the corresponding value of C_L. Repeat the same procedure to obtain a set of values as shown in Table 6.12.

Table 6.12: Arbitrarily chosen value of V_L, C_L from the graph

Sl. No.	V_L (m/s)	C_L (kg/m³)	C_V (kg/m³)	$1/C_L$ (m³/kg)	$1/C_V$ (m³/kg)	$(1/C_L - 1/C_V)$ (m³/kg)	$\frac{L_L \cdot C_L}{S} = \frac{V_L}{(1/C_L - 1/C_V)}$ (m/s)(kg/m³)
1	6×10^{-2}	28	531.85	0.0357	0.0013	0.0338	1.78
2	5×10^{-2}	34	531.85	0.0294	0.0019	0.0275	1.82
3	4×10^{-2}	38	531.85	0.0263	0.0019	0.0244	1.64
4	3×10^{-2}	42	531.85	0.0238	0.0019	0.0219	1.37
5	2×10^{-2}	46	531.85	0.0217	0.0019	0.0198	1.01
6	1×10^{-2}	63	531.85	0.0159	0.0019	0.0140	0.71

Using the arbitrarily obtained values of V_L, C_L and under flow concentration C_V, perform the calculation and get the corresponding value of $L_L \cdot C_L/S$. Then, plot the graph V_L vs $(L_L \cdot C_L/S)$ as shown in Fig. 6.20.

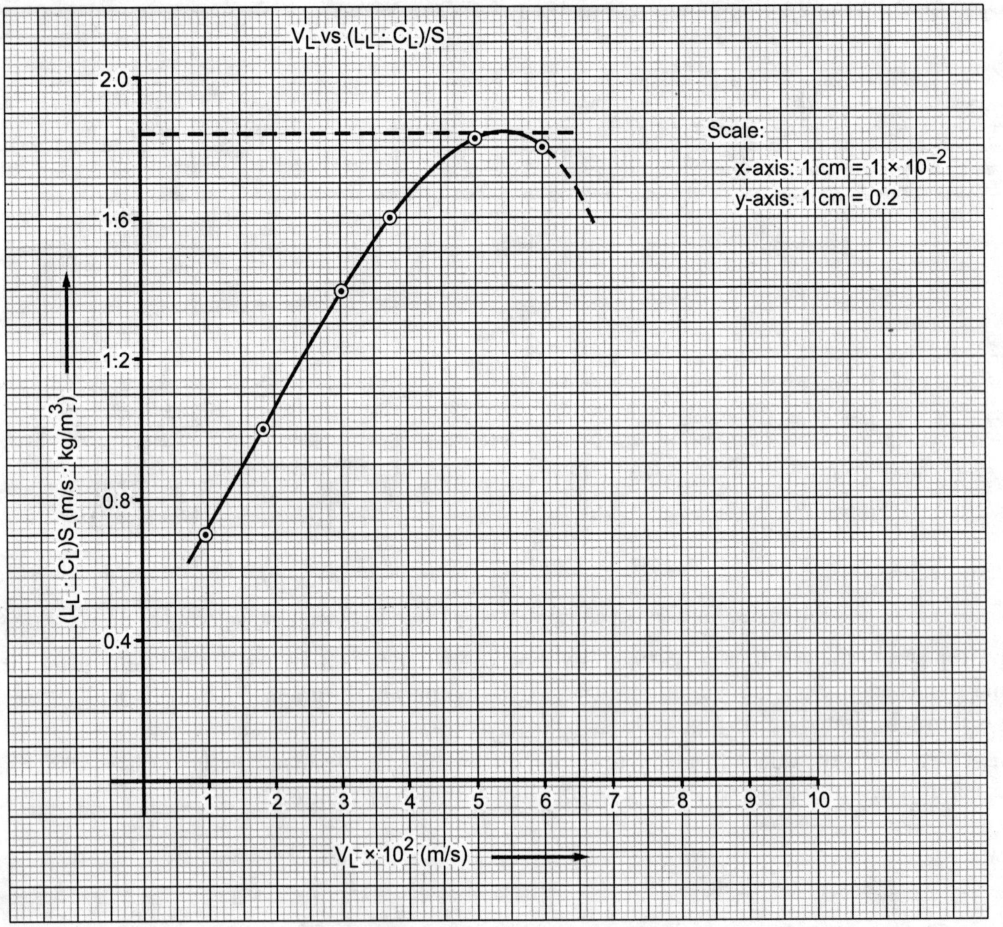

Fig. 6.20: Maximum value of $L_L \cdot C_L/S$

From the equation of maximum thickener area, we have

$$\frac{L_L \cdot C_L}{S} = \frac{V_L}{(1/C_L - 1/C_V)}$$

$$\frac{L_L \cdot C_L}{S} = 1.84 \text{ (m/s)(kg/m}^3\text{)}$$

$$S = \frac{4{,}00{,}000}{24 \times 60 \times 60} \times \frac{1}{1.84}$$

$$S = 2.516 \text{ m}^2$$

The maximum thickener area is 2.516 m^2. **Ans.**

REVIEW QUESTIONS

6.1 Define sedimentation.

6.2 Briefly explain the batch settling test.

6.3 What is the application of batch settling test. Explain.

6.4 What is Coe and Clevenger theory? Explain.

6.5 What is Kynch theory? Explain.

EXERCISES

6.1 Derive an equation to calculate the thickener area for a continuous thickener.

6.2 A single batch settling test was made on a liquid slurry. The interface between the liquid and suspension solids were observed as a function of time and the results are tabulated below. The test was made using 235 g of limestone per litre of slurry.

Sl. No.	Interface height (cm)	Time (hr)
1	36	0.00
2	32.4	0.25
3	28.6	0.50
4	21.0	1.00
5	14.7	1.75
6	12.3	3.00
7	11.6	4.75
8	9.8	12.00
9	8.8	20.00

Prepare a curve showing the relationship between settling rate and the solid concentration. Using this plot a curve of settling velocity vs solid concentration. Find the thickener area if the slurry fed at a rate of 30,000 kg dry solids per hr to produce a thickened sludge of 530 g limestone per litre.

6.3 150 tonnes/hr of suspension is to be concentrated from 165 g/l to 800 g/l in Dorr thickener. Determine the minimum thickener area of thickener using the following batch sedimentation test results. Specific gravity of solid is 2.46.

Sl. No.	Pulp height (cm)	Time (hr)
1	90	0.00
2	62	0.10
3	43	0.25
4	28	0.50
5	21	0.75
6	15	1.00
7	8	2.00
8	3	4.00

6.4 A 3% by weight $CaCO_3$ slurry was subjected to a batch sedimentation test. The density of solid was 2.63 g/cc and that of the liquid was 1 g/cc. The test results are indicated below:

Sl. No.	Z (m)	θ (min)
1	176	0
2	100	20
3	74	40
4	57	60
5	42	80
6	34	100

7	26	120
8	18	140

where, Z = Height of settling interface, m

θ = Settling time, min

Determine the maximum cross-sectional area of a continuous thickener that is to handle 8,00,000 kg of solids per day, if the initial solid concentration in the feed 3% by weight and the final concentration is to 40% by weight of solids.

6.5 The following data was obtained from a single batch sedimentation test on a slurry of initial concentration of 64.5 g/l. Density of solids is 2.5 g/cc and that of liquid is 1 g/cc. Determine the cross-sectional area of a continuous thickener to handle 100 tons/day of dry solids to give a final underflow concentration of 450 g/l.

Sl. No.	C_L (g/l)	V_L (cm/hr)
1	64.5	140
2	70.9	103
3	94.3	72
4	111.7	49
5	140.0	27
6	174.0	16
7	222.0	10
8	337.0	6.4

[VTU Exam, Sept. 1995]

6.6 Determine the minimum cross-sectional area of a continuous thickener required to handle 150 tonnes/day of dry solids from initial concentration of 5% to give an under-flow concentration of 45 by conducting a batch sedimentation test. The test data are given below.

Sl. No.	Interface height, Z (cm)	Time, θ (s)
1	31.5	0
2	30.5	29
3	29.5	55
4	28.5	78
5	27.5	107
6	26.5	131
7	25.5	156
8	24.5	182
9	23.5	212
10	22.5	239
11	21.5	265
12	20.5	295
13	19.5	325
14	18.5	349
15	17.5	375

16	16.5	404
17	15.5	436
18	14.5	460
19	13.5	456
20	12.5	522
21	11.5	550
22	10.5	583
23	9.5	616
24	8.5	653
25	7.5	734
26	6.5	871
27	5.5	1058
28	4.5	1321
29	3.5	1623
30	3.0	1864
31	2.5	2400

Density of $CaCO_3$ = 2560 kg/m^3

Density of water = 1000 kg/m^3

Initial height = 31.5 cm

6.7 A batch sedimentation test was carried out upon silica slurry with water. Initial concentration was 0.125 g/cc of slurry. The settling data is given below. Final height of settled solids is 5.53 cm. Determine the maximum cross-sectional area of continuous thickener to handle 150 tonnes/day of dry solids to a final concentration of 0.375 g/cc of slurry.

Sl. No.	Height, Z (cm)	Time, θ (min)
1	34	0
2	25	10.8
3	20	16.8
4	15	26.4
5	10	43.1
6	7.5	65.8
7	6.1	89.5
8	6.1	100.0
9	5.9	110.0

where, Z = Height of interface, cm

θ = Settling time, hr

Determine the maximum thickener area required.

❑❑❑

7 CHAPTER Filtration

7.1 INTRODUCTION

Separation of solids from suspension in liquid with the help of a porous medium or screen which retains the solids and allows the liquid to pass through termed as filtration.

Filtration is one of the most common applications of the flow of fluids through packed bed. As carried out industrially, it is exactly analogue to the filtration carried out in chemical laboratory through a filter paper in a funnel. The object is the separation of solids from the fluid in which it is carried. In every case, the separation is accomplished by foraging the fluid through a porous membrane the solids particles are trapped within the pores of the membrane and build up as a layer on the surface of this membrane. The fluid, may be either gas or liquid which passes through the bed of solids and through the retaining membrane.

The suspension to be filtered is known as slurry and the porous medium used to retain the solids is known as filter medium. The accumulated solids on the filter are referred to as filter cake, while the clear liquid passing through the filter is called filtrate.

When solids are present in liquid with very low concentration i.e., not exceeding 1% w/v, the process of its separation from liquid is called clarification. Clarification is generally employed when contaminating material is finely sub-divided, amorphous or colloidal in nature. These solids tend to plug the filter medium.

7.2 PROCESS OF FILTRATION

A typical filtration operation is shown in Fig. 7.1.

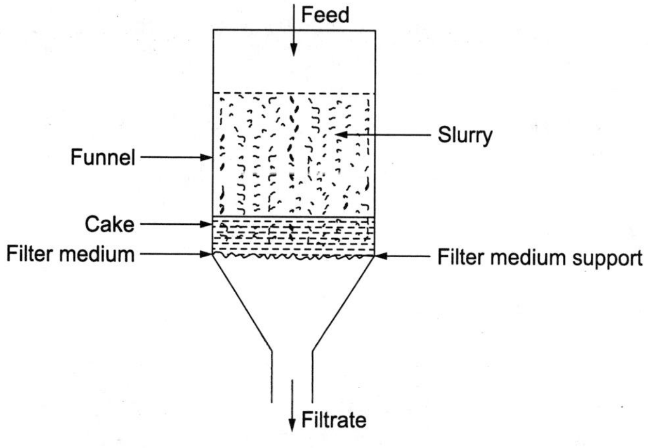

Fig. 7.1: Process of filtration

The process of the filter medium are smaller than the size of the particles to be separated. The filter medium either filter paper or muslin cloth is placed on the top of the support. When the slurry is pored over the filter medium, the fluid pass through the filter medium by virtue of pressure differential across the filter. Gravity is acting on the liquid column. Therefore, solids are trapped on the surface of the filter medium.

Once the preliminary layer of particle is deposited, further filtration is brought about wherein the filter medium serves only as a support. The filter will work efficiently after an initial deposit. In an industrial scale, large quantities of suspensions are mechanically handled. After a particular point of time the resistance offered by the filter cake is high that virtually stopped the filtration. For this reason, a positive pressure is applied on the filter cake or negative pressure is applied below the filter medium.

7.3 MECHANISM OF FILTRATION

The mechanism of filtration are as follows:

(*i*) Attractive force
(*ii*) Entanglement
(*iii*) Impingement
(*iv*) Straining

7.3.1 Attractive Force

The solids are retained on the filter medium as a results attractive forces between the particles and filter medium, as in the case of electrostatic precipitation.

7.3.2 Entanglement

Particles become entangled in the mass of fibres due to smaller size particles than the pore size. Thus the solids are retained on the filter medium.

7.3.3 Impingement

Solids having the momentum move along the path of streamline flow and strike the filter medium. Thus the solids are retained on the filter medium.

7.3.4 Straining

The particles of larger size cannot pass through the smaller pore size of the filter medium.

In practice, filtration may combine various mechanism.

7.4 TYPES OF FILTRATION

Based on the mechanism, filtration may be broadly classified into the types:

(*i*) Surface filtration
(*ii*) Depth filtration
(*iii*) Cake filtration

7.4.1 Surface Filtration

It is a screening action by which pores or holes of the filter medium prevent the passage of the solids. A typical surface filtration is shown in Fig. 7.2.

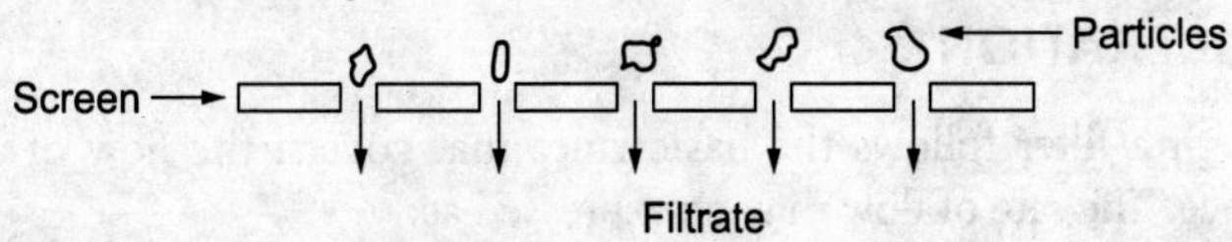

Fig. 7.2: Surface filtration

The mechanisms of straining and impingement are responsible for surface filtration. For this purpose plates with holes or woven sieves are used. Its efficiency is defined in terms of mean or maximum pore size.

7.4.2 Depth Filtration

In this process slurry penetrates to a point where the diameter of solids particle is greater than that of the tortuous void or channel. A typical depth filtration is shown in Fig. 7.3.

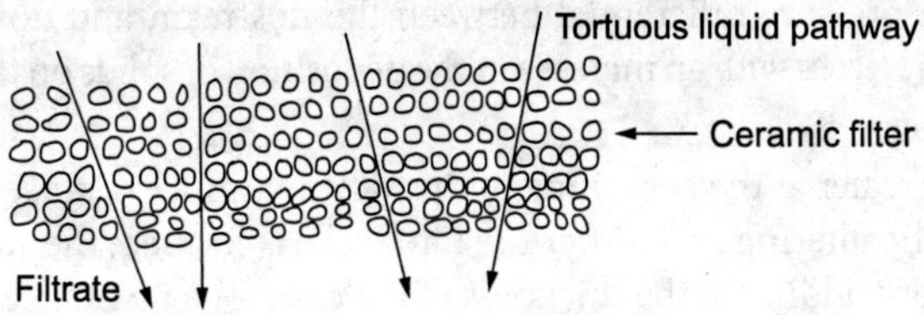

Fig. 7.3: Depth filtration

Depth filtration is aided by the mechanism of entanglement. The solids are retained with a gradient density structure by physical restriction or by adsorption properties of the medium. It is extensively used for the removal of small amount of contaminants from relatively large volume of liquids. These are also used for roughing or prefiltering solutions.

For example, ceramic filter and sintered bed filter. A special case of depth filtration is the sheet filtration, in which the slurry is passed through a series of pads mounted endwise to the flow of fluids. The pads commonly consists of a mixture of asbestos fibers embedded in cellulose.

7.4.3 Cake Filtration

A cake filter consists of a coarse woven cloth through which a concentrated suspension of rigid particle is passed so that they bridge the holes and form a bed.

The cloth may be in effective when the holes are likely to be much larger than the particles to be removed. Therefore, the cloth is presented with a concentrated suspension of rigid particles that bridge the holes and form a bed. The particles will be held back, while liquid passes through the small interface. A typical cake filtration is shown in Fig. 7.4.

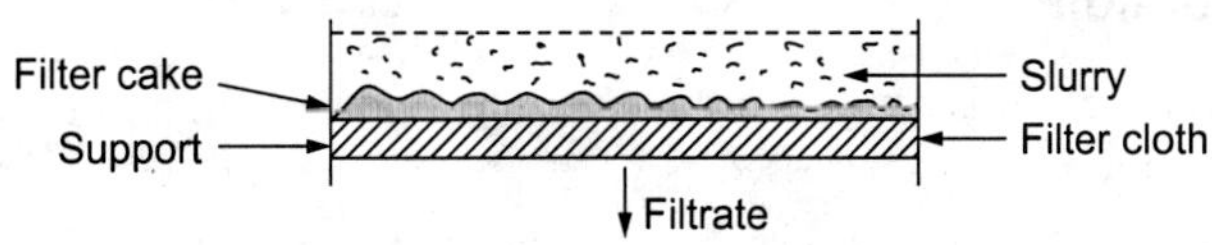

Fig. 7.4: Cake filtration

For example, the diatomic cake filter. This cake can remove sub-micrometer colloidal particles with high efficiency.

7.5 THEORY OF FILTRATION

The flow of liquid through a filter follows the basic rules that govern the flow of any liquid through the medium offering resistance. The rate of flow may be expressed as:

$$\text{Rate} = \frac{\text{Driving force}}{\text{Resistance}} \quad ...(7.1)$$

The rate of filtration may be expressed as:

$$\text{Rate} = \frac{dV}{dt} \quad ...(7.2)$$

where, V = Volume, m^3

t = Time, hr

The driving force is the pressure differential between the upstream and downstream of the filter. The resistance is not constant. It increases with an increase in the deposition of solids on the filter medium. Therefore, filtration is not steady state.

The rate of flow will be greater at the beginning of the filtration process, since the resistance is minimum and will be proportional to the filtering area. After the filter cake is formed, the surface acts as filter medium and solids continuously deposit adding to the thickness of the cake. The resistance to flow is related to several factors. This phenomena is explained by the Fig. 7.5.

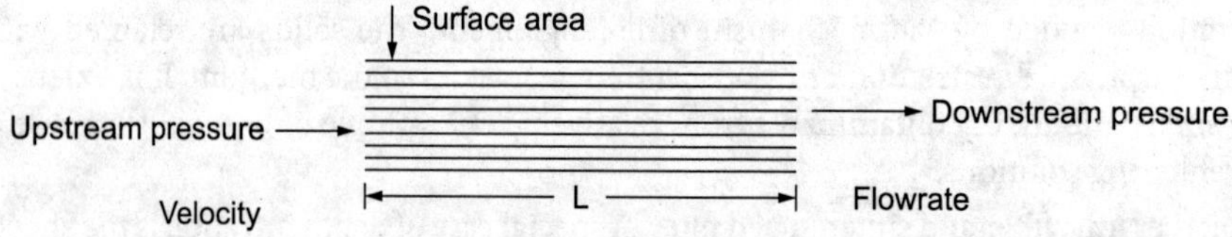

Fig. 7.5: Grannule bed act as a bundle of capillaries

The resistance to flow may be expressed as:

$$\text{Resistance to movement} = \frac{\text{Upstream pressure} - \text{Downstream pressure}}{\text{Length of capillaries}} \quad ...(7.2)$$

The flow rate of the fluid through the filtration chamber may be determined using the following equations:

(*i*) Poiseuille's equation

(*ii*) Darcy's equation

(*iii*) Kozeny-Carman equation

7.5.1 Poiseuille's Equation

Poiseuilles considered that filtration is similar to the streamline flow of a liquid under pressure through the capillaries. Poiseuillies equation may be expressed as:

$$V = \frac{\pi \Delta P r^4}{8L\eta} \quad ...(7.3)$$

where, V = Flow rate of liquid through the filtration chamber, m^3/s

ΔP = Pressure difference across the filter, N/m^2

r = Radius of the capillary in the bed, m

L = Thickness of the filter cake, m

η = Viscosity of the filtrate $N.s/m^2$

If the cake is composed of a bulky mass of particles and the liquid flows through the interstices, then the flow of liquids through these may be expressed by the Poiseuille's equation.

7.5.2 Darcy's Equation

Poiseuille's law assumed that the capillary found in the filter are highly irregular and non-uniform. Therefore, if the length of a capillary is taken as the thickness of the bed, a correction factor for radius is applied so that the rate equation is closely approximated and simplified. The factors influencing the rate of filtration has been incorporated into an equation by Darcy. It may be expressed as:

$$V = \frac{KA\Delta P}{\eta L} \qquad ...(7.4)$$

where,
K = Permeability co-efficient of the cake, m^2
A = Surface area of the porous bed, m^2
V = Rate of filtrate, m^3/s
ΔP = Pressure difference across the filter, N/m^2
η = Viscosity of the filtrate, $N.s/m^2$
L = Thickness of the filter cake, m

The term K, permeability co-efficient depends on the characteristics of the cake, such as porosity, specific surface area of the cake and compressibility factor. The permeability co-efficient of the cake is the flow rate of a liquid of a unit viscosity across a unit area of cake having unit thickness under a pressure gradient of unity.

The model relates not only to filter bed or cakes but also applies to depth filter. Darcy's equation is valid for liquids flowing through sand, glass beads and various porous media.

7.5.3 Kozency-Carman Equation

Poiseuille's equation is made applicable to porous bed based on the capillary type structure by including additional parameters. Thus, the modified equation which is widely used for filtration is known as Kozeny Carman equation. It may be expressed as:

$$V = \frac{A}{\eta S^2} \times \frac{\Delta P}{KL} \times \frac{\varepsilon^3}{(1-\varepsilon)^2} \qquad ...(7.5)$$

where,
V = Rate of flow, m^3/s
A = Surface area of the porous bed, m^2
ΔP = Pressure difference across the filter, N/m^2
ε = Porosity of the bed
η = Viscosity of the filtrate, $N.m/s^2$
S = Specific surface area of the particles, m^2/kg
K = Kozeny constant
L = Thickness of the filter cake, m

The Kozeny's constant is usually taken as 5. The effect of compressibility of the cake on flow rate can be appreciated from equation (7.1), since the flow rate is proportional to $\varepsilon^3/(1 - \varepsilon)^2$. A 10% change in porosity can produce almost 3 fold change in volume.

Limitation of Kozeny Carman equation: Kozeny-Carman equation does not take into account of the fact that the depth of the granular bed is lesser than the actual path traversed by the fluid. The actual path is not straight throughout the bed, but it is sinuous or tortuous

7.6 MODIFICATION OF KOZENY-CARMAN EQUATION

In all most all practical cases of filtration flow is laminar and represented by Kozeny-Carman equation.

$$\frac{\varnothing P}{L} = \frac{180(1-\varepsilon)^2 \mu V_s}{\varepsilon^2 D_P^2} \quad ...(7.6)$$

This equation relates the pressure drop through the cakes to the flow rate, the cake porosity, thickness and solid particle diameter. Some modification of the equation is necessary so that the measurable variable of filtration can be introduced into it. Kozeny-Carman equation can be written in terms of specific surface area of the particles in more useful form.

The specific surface area of the particle related to its diameter by the expression

$$D_P = \frac{6}{S_P} \quad ...(7.7)$$

Substituting equation (7.7) in equation (7.6), we get

$$\frac{\varnothing P}{L} = \frac{5(1-\varepsilon)^2 \mu V_s S_P^2}{\varepsilon^3} \quad ...(7.8)$$

Solving this equation for velocity of flow, we get

$$\frac{1}{A}\frac{dV}{dt} = \frac{\Delta P \varepsilon^3}{5L(1-\varepsilon)^2 \mu S_P^2} \quad ...(7.9)$$

The cake thickness may be related to the volume of filtrate by material balance, since thickness will be proportional to the volume of the feed delivered to the filter.

$$LA(1-\varepsilon)\rho_s = W(V + \varepsilon LA) \quad ...(7.10)$$

The final term of equation (7.10) represent the volume of filtrate held in the filter cake. It may be negligible.

$$LA(1-\varepsilon)\rho_s = WV \quad ...(7.11)$$

Substituting equation (7.11) in equation (7.8), we get

$$\frac{1}{A}\frac{dV}{dt} = \frac{\Delta P \varepsilon^3}{5WV(1-\varepsilon)\mu S_P^2 / A\rho_s} \quad ...(7.12)$$

$$\Rightarrow \quad \frac{1}{A}\frac{dV}{dt} = \frac{\Delta P}{\{5(1-\varepsilon)S_P^2/\rho_s \varepsilon^3\}(\mu VW/A)} \quad ...(7.13)$$

$$\Rightarrow \quad \frac{1}{A}\frac{dV}{dt} = \frac{\Delta P}{\alpha(\mu VW/A} \quad ...(7.14)$$

where, $\alpha = 5(1-\varepsilon)S_P^2/\varepsilon^3\rho_s)$

S_P = Specific surface area, m^2/m^3

V = Volume of filtrate, m^3

ρ_s = Density of solid in the cake

W = Weight of solid in the feed slurry per volume of the liquid, kg/m^3

V_s = Velocity of flow, m/s

α = Specific cake resistance

Considering equation (7.10), we have

$$LA(1-\varepsilon)\rho_s = W(V + \varepsilon LA) \quad ...(7.10)$$

$$L(1-\varepsilon) = W(V + \varepsilon LA)/A\rho_s \quad ...(7.15)$$

Substituting equation (7.15) in equation (7.9), we get

$$\frac{1}{A}\frac{dV}{dt} = \frac{\Delta P}{5\{(1-\varepsilon)S_P^2\mu/\varepsilon^3\rho_s\}(WV+\varepsilon WLA)/A} \quad ...(7.16)$$

$$\Rightarrow \quad \frac{1}{A}\frac{dV}{dt} = \frac{\Delta P}{\alpha\mu(WV+W\varepsilon LA)/A}$$

$$\Rightarrow \quad \frac{1}{A}\frac{dV}{dt} = \frac{\Delta P}{(\alpha\mu WV/A)+(\alpha\mu W\varepsilon/q)}$$

$$\Rightarrow \quad \frac{1}{A}\frac{dV}{dt} = \frac{\Delta P}{\mu(\alpha WV/A)+\mu R_m} \quad ...(7.17)$$

This is the modified Kozeny-Carman equation, which used in filtration.

The process of filtration may be carried out by two method:

(*i*) Constant pressure filtration

(*ii*) Constant rate filtration

7.6.1 Constant Pressure Filtration

The method in which the pressure drop over the filter is constant throughout the runs so that the rate of filtration is maximum, at the start of filtration and decreases continuously towards the end of the run is called constant pressure filtration. It is carried out by applying certain pressure at the inlet and maintaining it constant throughout the run. The rate of filtration is expressed by the equation.

$$\frac{1}{A}\frac{dV}{dt} = \frac{\Delta P}{\mu(\alpha WV/A+R_m)} \quad ...(7.17)$$

where,
V = Volume of the filtrate, m^3
A = Surface area of the filter, m^2
t = Time, hr
ΔP = Pressure difference, N/m^2
μ = Viscosity of the filtrate, $N.s/m^2$
α = Specific cake resistance, m/kg
W = Weight of dry cake, kg
R_m = Filter medium resistance, m^{-1}

Inversing equation (7.17), we get

$$\frac{Adt}{dV} = \frac{\mu(\alpha WV/A+R_m)}{\Delta P} \quad ...(7.18)$$

$$\Rightarrow \quad \frac{dt}{dV} = \frac{\mu}{A\Delta P}[\alpha WV/A+R_m]$$

$$\Rightarrow \quad dt = \frac{\mu}{A\Delta P}[\alpha WV/A+R_m]dV \quad ...(7.19)$$

Integrating equation (7.19) between the limit, we get

$$\int_0^t dt = \int_0^V\left(\frac{\mu\alpha W}{A^2\Delta P}\right)VdV+\int_0^V\left(\frac{\mu R_m}{A\Delta P}\right)dV$$

$$\Rightarrow \quad t = \frac{\mu\alpha W}{A^2\Delta P}\frac{V^2}{2} + \frac{\mu R_m}{A\Delta P}V \qquad ...(7.20)$$

$$\Rightarrow \quad \frac{t}{V} = \frac{\mu\alpha WV}{2A^2\Delta P} + \frac{\mu R_m}{A\Delta P} \qquad ...(7.21)$$

$$\Rightarrow \quad \frac{t}{V} = \frac{k_c \cdot V}{2} + \frac{1}{q_0} \qquad ...(7.22)$$

where,
$$k_c = \frac{\mu\alpha W}{A^2\Delta P} \qquad ...(7.23)$$

$$q_0 = \frac{A\Delta P}{\mu R_m} \qquad ...(7.24)$$

By plotting t/V vs V, the specific cake resistance 'α' and filter medium resistance 'R_m' can be determined. The approximate value of specific cake resistance is 1.91×10^{11} m/kg and that of the filter medium resistance is 2.65×10^{10} m^{-1}.

7.6.2 Constant Rate Filtration

The method in which pressure drop is varied usually from minimum at the start of the filtration and maximum at the end of the filtration. So that the rate of filtration is constant throughout the run is called constant rate of filtration.

In constant rate of filtration starts at low inlet pressure and continuously increase the pressure to overcome the resistance of the cake until the maximum pressure is reached towards the end of the run maintain the rate.

From the rate of filtration, we have

$$\frac{1}{A}\frac{dV}{dt} = \frac{\Delta P}{\mu(\alpha WV/A + R_m)} \qquad ...(7.17)$$

Neglecting the filter medium resistance, the equation (7.17) reduced to

$$\frac{1}{A}\frac{dV}{dt} = \frac{\Delta PA}{\mu(\alpha WV/A)} \qquad ...(7.25)$$

When the filtrate flow rate is constant

$$\frac{V}{At} = \frac{\Delta PA}{\mu(\alpha WV)} \qquad ...(7.29)$$

$$\Rightarrow \quad \frac{V}{t} = \frac{\Delta PA^2}{\mu(\alpha WV)} \qquad ...(7.30)$$

where,
$\frac{V}{t}$ = Constant rate filtration, m^3/s
ΔP = Pressure difference, N/m^2
A = Filtration area, m^2
μ = Viscosity of the liquid, kg/m.s
α = Specific cake resistance, m/kg
W = Weight of the solid in the feed slurry per volume of the liquid, kg/m^3
V = Volume of filtrate, m^3

7.7 FACTORS INFLUENCING THE FILTRATION

The list of the factors which influence the filtration are as follows:

(*i*) Surface area of the filter medium
(*ii*) Pressure drop across the filter medium
(*iii*) Viscosity of the filtrate.

7.7.1 Surface Area of the Filter Medium

The rate of flow of filtrate flowing through the filter is inversely proportional to the specific surface area of the filter bed. On the other hand, if the surface area of the filter medium is considered, the rate of filtration is directly proportional to the surface area of the filter medium. Hence, the rate can be increased either using a large filter or connecting a number of small filter in parallel.

7.7.2 Pressure Drop Across the Filter Medium

The rate of filtration is directly proportional to the overall pressure drop across the filter medium and filter cake. The pressure drop can be achieved in a number of ways:

(*i*) Gravity
(*ii*) Applying pressure
(*iii*) Reducing pressure
(*iv*) Centrifugal force

7.7.2.1 Gravity

A simple method of obtaining the pressure difference is by maintaining a head of slurry above the filter medium. The pressure developed depends on the density of the slurry.

7.7.2.2 Applying Pressure

The most common method of obtaining pressure difference is by applying pressure on the surface of the slurry, i.e., pumping the slurry on to the filter. The industrial plant may be operated at a pressure upto 1500 kPa. However, in the early stages of filtration, pressure difference should be less. This is to prevent the pores of the filter medium from clogging, which subsequently increases resistance to flow.

7.7.2.3 Reducing Pressure

The pressure underneath the filter medium may be reduced below atmospheric pressure by connecting the filtrate receiver to a vacuum pump. This create pressure difference across the filter. This has limited application because the pressure difference of about 100 kPa may be achieved. The reduction of pressure lower the boiling point of liquid so that it is possible for the filtrate to boil in the receiver. One dis-advantage of reducing pressure is that apart from the loss of liquid, the vapour may damage the vacuum pump.

7.7.2.4 Centrifugal Force

It can replace the gravitational force and is used to replace the rate of filtration.

7.7.3 Viscosity of the Filtrate

Rate of filtration is inversely proportional to the viscosity of the fluid. The viscosity of the liquid is important, since the resistance to flow occurs as the filtrate flows through the filter cake.

Raising the temperature of the liquid lower the viscosity, which in terms increase the rate of filtration. This is not practible if filtrate is more volatile.

Another alternative method to decrease the viscosity is to dilute the solution to be filter.

In brief, surface area and pressure difference are the two factors, which are applied in industrial practice.

7.8 FILTER MEDIA

The filter media acts as a mechanical support for the filter cake and is also responsible for collection of solids. The characteristics of filter media are as follows:

(*i*) It should have sufficient mechanical strength

(*ii*) It must be inert

(*iii*) It should not absorb the dissolved materials.

(*iv*) It should have maximum passage for the liquid, while retaining the solid.

The magnitude of the resistance of the filter medium will change due to the layers of solids deposited, which may block the pores over the entrance of the channel. Therefore, pressure should be maintained low at the beginning to avoid the plugging of the pores. The usual procedure is to filter at constant rate by increasing the pressure as necessary.

On continued filtration, the thickness of the cake further build up, hence the rate of filtration decreases. When the rate is uneconomical the filtration is topped. The filter cake is removed and the process of filtration can be restarted. The following materials are used as filter media.

(*i*) Woven material such as felt or cloth.

(*ii*) Perforated sheet metal.

(*iii*) Bed of granular solid built up on a supporting medium.

(*iv*) Membrane filter media

(*v*) Surface type centrifuge

(*vi*) Depth type centrifuge

7.9 FILTER AIDS

It is an agent consisting of solids particles that improves filtering efficiency and that is either added to the suspension to be filtered or placed on the filter as a layer through which the liquid must pass.

Filter aid forms a surface deposit which screen out the solids and also prevent the plugging of the supporting filter medium. The characteristics of the filter aids are:

(*i*) It should be chemically inert

(*ii*) It should be free from impurities

(*iii*) It should have low specific gravity

(*iv*) It should be recoverable.

Example of filter aid are:

(*i*) Keiselguhr

(*ii*) Charcoal

(*iii*) Talc

(*iv*) Bentonite

(*v*) Fullers earth

7.10 APPLICATION OF FILTRATION

The application of filtration are as follows:

(*i*) The coffee filter to separate the coffee infusion from the ground.
(*ii*) HEPA filters in air conditioning to remove particles from the air
(*iii*) Belt filters to extract precious metals in mining.
(*iv*) Vertical plate filter such as those used in crowe process.
(*v*) Nutsche filters typically used in pharmaceutical applications
(*vi*) In the laboratory Buchner funnel is often used, with a filter paper serving as the porous barrier.
(*vii*) Air filters are commonly used to remove air borne particulate matter
(*viii*) Oil filter in automobiles, often as canister or cartridge
(*ix*) Aquarium filter
(*x*) Carbon dioxide removal from breathing gas
(*xi*) Activated carbon filter

7.11 CLASSIFICATION OF FILTRATION EQUIPMENT

The filtration equipment may be classified by various manner. They are as follows:

Classification of filter based on the application of external force:

(*i*) **Pressure filter:** Plate and frame filter press, meta filter.
(*ii*) **Vacuum filter:** Leaf filter, Buchner funnel
(*iii*) **Centrifugal filter:** Perforated basket centrifuge.

Classification of filter based on the operation of filter:

(*i*) Continuous filtration
(*ii*) Discontinuous filtration

Classification of filter based on the nature of filtration:

(*i*) Cake filters
(*ii*) Clarifying filters
(*iii*) Cross flow filters

7.12 INDUSTRIAL FILTERS

There are many industrial filters. But, only a few are discussed here:

(*i*) Sand filter
(*ii*) Plate and frame filter press
(*iii*) Leaf filter
(*iv*) Rotary drum filter
(*v*) Horizontal belt filter
(*vi*) Bag filter
(*vii*) Meta filter
(*viii*) Suspended batch centrifuge

7.12.1 Sand Filter

The simplest possible filter is a wooden box with a perforated bottom filled with loose sand. The filter of exactly this type have often been used, are quite completely standardized. It is useful mainly where relatively all amounts of solids are to be removed from the liquid and where very large volume of liquid to be handled at the minimum cost. A typical sand filter is shown in Fig. 7.6.

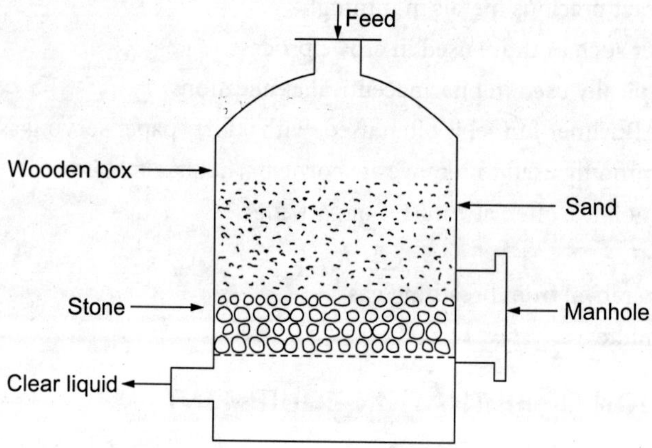

Fig. 7.6: Sand filter

The sand filter is used for the filtration of boiler feed water or for similar purposes.

The bottom of the tank either mounted on a false bottom or connected to a manifold concrete stainer. These are made of brass and have narrow slot bed in them. Over the strainers a layer of several inches of moderately coarse gravel and top of that is the sand that act as the actual filter medium. The layer of sand may be 2 m deep.

The water to be filtered is introduced at the top on to baffle which prevents curving the sand by a direct stream. The clean water is drawn off through the strainers at the bottom. When the precipitate clogged the sand to such an extent that the flow is retarded, it removed by back washing. The operation consists of introducing waste water through the inlet so that it may flow through the sand bed and out through the connection that is finally the outlet. The sand filters of this type are applicable only to the separation of precipitates that can be removed from the sand in this manner and that are to be retarded.

7.12.2 Plate and Frame Filter Press

The mechanism involved in plate and frame filter press is surface filtration. The slurry enters in to the frame by pressure and flow through the filter medium. The filtrate is collected on the surface and send to the outlet. A number of plate and frame are used so that surface are increases and consequently large volume of slurry can be processed simultaneously with or without washing. A typical plate and frame filter press is shown in Fig. 7.7.

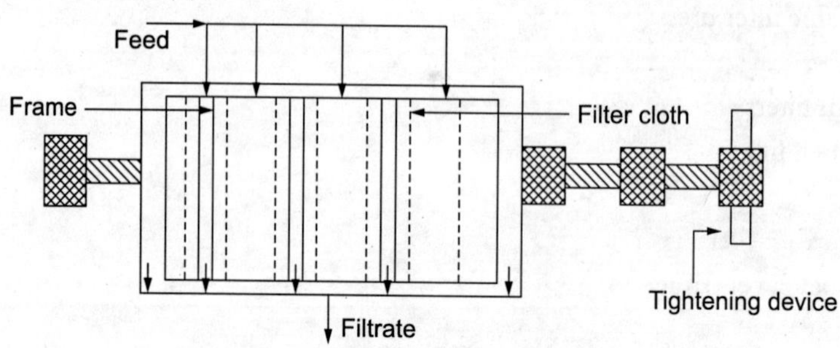

Fig. 7.7: Plate and frame filter press

The filter press is made of two unit, plate and frame. These are usually made of Aluminium alloy. These are also lacquered for protection against corrosive chemicals and suitable for steam sterilization.

It contains an open space inside wherein the slurry reservoir is maintained for filtration and opening to receive the slurry. The number of plates and frames are employed so that filtration area is as large as necessary, in other words, a number of filtration units are operated in parallel. Channel for the slurry inlet and filtrate outlet can be arranged by fitting eyes to the plate and frames, these joints together to form a channel.

The working of plate and frame filter press involve two steps, namely filtration and washing of the cake. Slurry enters the frame from the feed channel and passes through the filter medium on to the surface of the plate. The solids form a filter cake and remain inside the frame.

The thickness of the cake is half the frame thickness, because on each side of the frame filtration occurs. Thus two filter cakes are formed which meet eventually in the centre of the frame. In general, there will be an optimum thickness of the filter cake for any slurry, depending on the solid content in the slurry and the resistance of the filter cake.

The filtrate drains between the projection on the surface of the plate and escapes from the outlet. As the filtration proceeds the resistance of the cake increases and filtration rate decreases. At a certain point it is preferable to stop the process rather than continuing at very low flow rate. The press is emplied and cycle is restarted. Necessary to wash the filter cake, the ordinary plate and frame filter press is unsatisfactory.

7.12.3 Leaf Filter

It consists of a narrow frame enclosing a drainage screen or grooved plate. The frame may be of any shape, circular, square or rectangular. The whole unit is covered with filter cloth. The outlet for the filtrate connects to the interior of the frame through suction. A typical leaf filter is shown in Fig. 7.8.

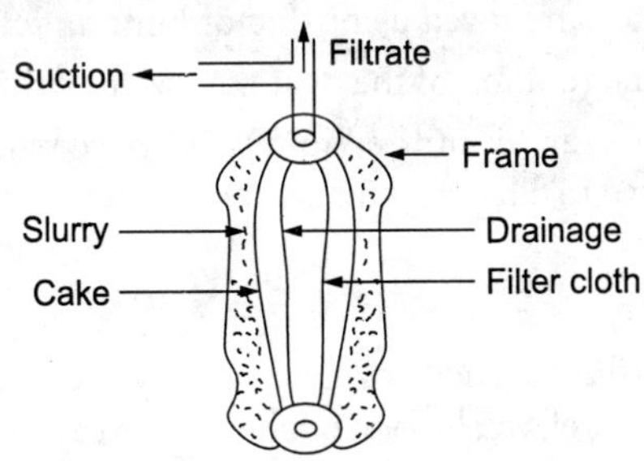

Fig. 7.8: Leaf filter

The filter leaf is immersed in the slurry. The vacuum system is connected to the filtrate outlet. The slurry passes through the filter cloth. Finally filtrate enters the drainage canal and goes through the outlet into the receiver. Air is passed to flow in reverse direction, which facilitates removal of cake.

The leaf filter is satisfactory, if the solid content of the slurry is not too high, about 5% of the total suspension. Leaf filter is a versatile piece of equipment. It is probably the oldest form of filter used for batch processes. A number of units can be connected in parallel to increase the surface area for filtration.

7.12.4 Rotary Drum Filter

It works on the principle of filtering the slurry through sieve like mechanism on a rotating drum surface.

The rotary drum filter consists of a metal cylinder mounted horizontally. The drum may be upto 3 m in diameter and 3.5 m in length and surface area of about 20 m^2. The curved surface is a perforated plate, which support a filter cloth. The drum is radically portioned dividing annular space into separate compartments. Edge of it is connected by an internal pipe to centre of the drum through a rotating valve. A typical rotating drum filter is shown in Fig. 7.9.

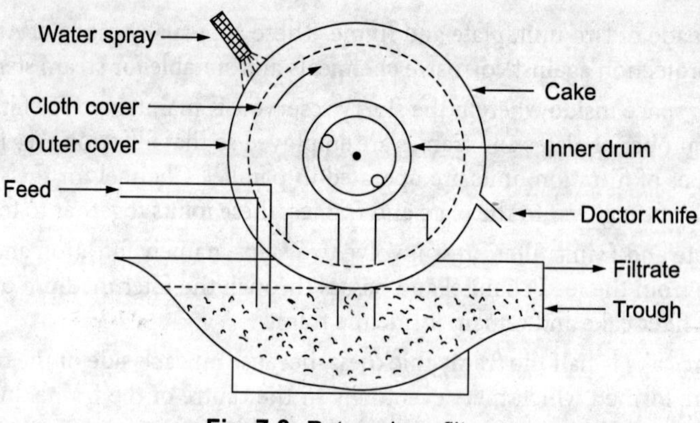

Fig. 7.9: Rotary drum filter

The drum is rotated at a speed less than one rotation per minute. The drum just enters the slurry in the trough. As it dips, vacuum is applied in the segment so that the solid is build up on the surface. The liquid passes through the filter cloth into an internal pipe and valve. Finally, the filtrate reaches the collecting tank.

As the drum leaves the slurry section, it enters the drainage zone. Here, the excess of liquid is drawn inside. Special cake compression roller may be included at this stage so that the cake is consolidated by the compression of the cake. This improve the efficiency of washing and drying process.

As the drum leaves the drainage zone it enters the water wash section. Water is sprayed on the cake. A separate system of vacuum is applied on the panel in order to suck the wash liquid and air through the cake of the solids. Wash liquid is drawn through the filter into a separate collecting tank.

Then, the cake enters the drying zone, where hot air is blown on the cake. The cake may have the moisture content less than 1%. Finally, the cake is removed using doctor knife and discharged.

All the steps are completed in one rotation of the drum. Now, the drum is ready to receive a fresh slurry.

It is used for continuous operation and is utilized to filter slurry containing high concentration of solids upto 15-30%. Rotary drum filter is used to filter calcium carbonate and magnesium carbonate from its solution.

7.12.5 Horizontal Belt Filter

When the feed contain coarse fast settling particles of solid a rotary drum filter works poorly or not at all. The coarse particles cannot be suspended well in slurry trough and the cake that forms often will not adhere to the surface of the filter. In this situation a top feed horizontal belt filter may be used. A typical horizontal belt filter is shown in Fig. 7.10.

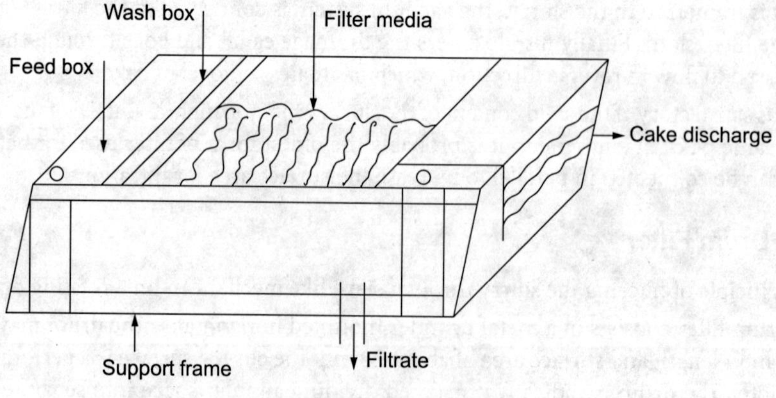

Fig. 7.10: Horizontal belt filter

Resemble a belt conveyor with a transversely rigid support or drainage belt carrying the cloth which is also in the form of endless belt. Central opening in the drainage belt over a longitudinal vacuum box into which the filtrate is drawn. Feed slurry flows onto the belt from a distributor at one end of the unit. After filtration the cake is discharged from the end.

This type of filters are especially useful in waste water treatment, since waste water often contains very wide range of particle size. They are available in the size from 0.5 to 6 m wide and 5 to 30 m long with filtration area upto 110 m^2. In this filter the vacuum is intermittent cut off and re-applied.

The belt is moved forward one half or a meter or so when the vacuum is off and is held stationary while vacuum is applied. This avoid difficulty of maintaining a good vacuum seal between the vacuum box and the moving filter media.

7.12.6 Bag Filter

It consists of a number of cylindrical fabric bag suspended in a sheet metal container. A typical bag filter is shown in Fig. 7.11.

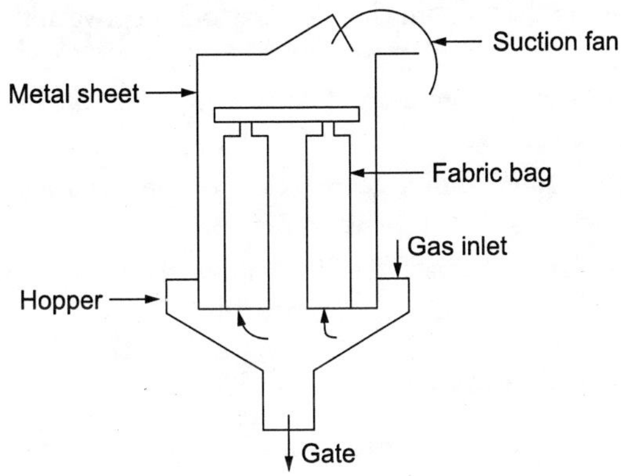

Fig. 7.11: Bag filter

During ordinary operation the gas to be filtered enters the hopper, passes up inside the bags through the fabric and move out from the top of casing to a main leading suction fan.

The suction unit is so designed at the whole apparatus works at less than atmospheric pressure during the filtering operation. The bags are suspended so that the air can be drawn easily. Since all the units are under diminished pressure, this casing and to pass into the bags, assisting the displacement of the dust. Such dust laden air passes into the hopper and hence to the other units to be filtered.

The greater portion of the dust falls into the hopper from which it is removed at interval from the gate. Then suction is stopped and vibration to be started. By this action particles get collected in the bag falls immediately to the hopper and then can be removed from the gate.

7.12.7 Meta Filter

It is an industrially used filter. For some manufacturing process the filter press is not convenient, when it is necessary to change from one preparation to another, because generally it is adopted to filter of large volume and the batches are operate in size, the volume of the pump and press may be large in proportion to batch, in cases Meta filter is adopted. A typical Meta filter is shown in Fig. 7.12.

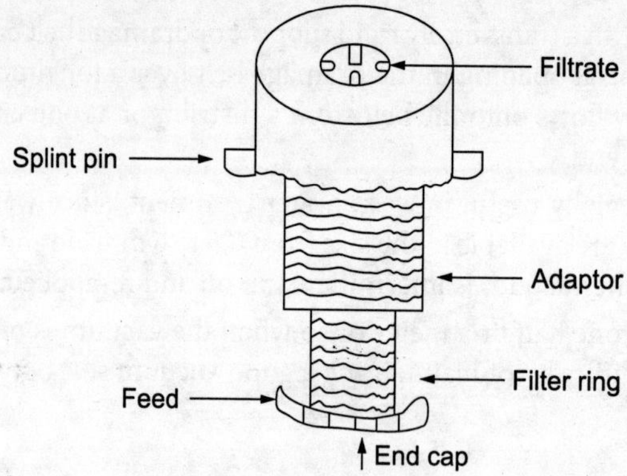

Fig. 7.12: Meta filter

A permanent filter is made up of rings on a grooved column. The rings 25 mm in external diameter and have raised proportion from 0.2 mm to suit requirement. The raised parts are only on one side of the rings, so that when the rings are threaded on the column channels are found between the raised parts of the ring and the flat side of the next.

The column of ring are used as unit to build up a filter in which they are permanent base in which a filter bed of Kieselguhr, charcoal, declorizing carbon or other filter aid.

The pressure filter may also be fitted with a steam jacket for the filtration of very viscous liquid such as syrup or oil, which can be rendered less viscous by warming.

This type filter can easily be cleaned by back flushing with water or steam, which is because of the shape of the pores in the rings will wash away the filter bed completely.

7.12.8 Suspended Batch Centrifuge

A common type of batch centrifuge in industrial processing is the top suspended centrifuge.

The perforated basket centrifuge ranges from 750 to 1200 mm in diameter and 450 to 750 mm in deep. The basket is held at the lower end of the free swinging vertical shaft driven from the top. A filter medium lies the perforated wall of the basket. A typical suspended batch centrifuge is shown in Fig. 7.13.

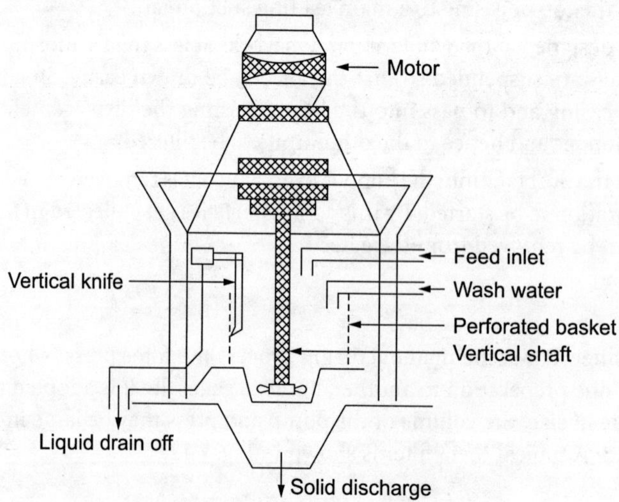

Fig. 7.13: Suspended batch centrifuge

Feed slurry enters the rotating basket through an inlet pipe. Liquor drains through the filter medium into the casing and out through a discharge pipe. The solids form cake 50 to 150 mm thick inside the basket. Wash liquid may be sprayed over the solid to remove soluble material. The cake is then spun to as dry as possible, sometimes at an higher speed than during the loading and washing steps. The motor is shut off and basket nearly stopped by means of a brake.

If the basket slowly turning at perhaps 30 to 50 rpm, cutting them out with discharges the solids by an unloaded knife, which peels the cake of the filter medium and drops it through an opening, in the basket floor. The filter medium is rinsed clean, the motor turned on and the cycle repeated.

The top suspended basket centrifuge is used extensively in sugar refining where they operate on about cycles 2 to 3 mintute per load and produce upto 5 tonne/hr of crystal per machine. Automatic controls are often provided for some of or all the steps in the cycle. In most of the cases where large tonnage of crystal is separated, however other automatic or continuous conveyor centrifuge are used.

The other form of batch centrifuge is driven from the bottom with the drive motor, basket and bag all suspended from vertical legs mounted on a base plate. Solids are unloaded by hand through top of the casing or plowed out through openings in the floor of the basket as top suspended machine.

Example 7.1. Slurry in water at 25°C are reported as follows at a constant pressure of 46.5 kPa. The area of the plate and frame filter press is 0.045 m^2 and the slurry concentration is 13.1 kg of solid/m^3 of filtrate. Calculate the specific cake resistance 'α' and the filter medium resistance 'R_m'.

Data:

Sl. No.	V (m^3)	t (s)
1	0.5×10^{-3}	17.5
2	1.0×10^{-3}	41.5
3	1.5×10^{-3}	72.5
4	2.0×10^{-3}	108.5
5	2.5×10^{-3}	152.0
6	3.0×10^{-3}	201.5

Solution:

$$T = 25°C$$

$$\Delta P = 46.5 \text{ kPa}$$

$$A = 0.045 \text{ m}^2$$

$$W = 13.1 \text{ kg of solid/m}^3 \text{ of filtrate}$$

$$\mu = 0.89 \times 10^{-3} \text{ kg/m.s}$$

Table 7.1: Plate and frame filter press test data

Sl. No.	V (m^3)	t (s)
1	0.5×10^{-3}	17.5
2	1.0×10^{-3}	41.5
3	1.5×10^{-3}	72.5
4	2.0×10^{-3}	108.5
5	2.5×10^{-3}	152.0
6	3.0×10^{-3}	201.5

Using the plate and frame filter press test data given in Table 7.1 perform the calculation and get corresponding value of t/V as shown in Table 7.2.

Table 7.2: Plate and frame filter press test data

Sl. No.	V (m³)	t (s)	t/V (s/m³)
1	0.5×10^{-3}	17.5	35,000
2	1.0×10^{-3}	41.5	41,500
3	1.5×10^{-3}	72.5	48,333
4	2.0×10^{-3}	108.5	54,256
5	2.5×10^{-3}	152.0	60,800
6	3.0×10^{-3}	201.5	67,167

Then, plot the graph t/V vs V as shown in Fig. 7.14.

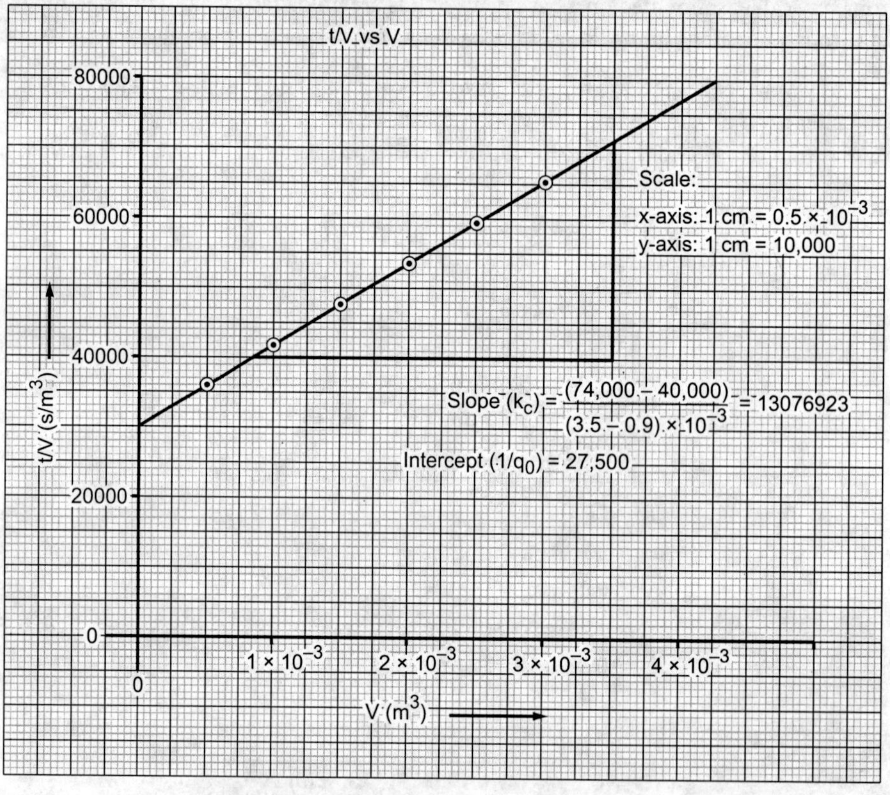

Fig. 7.14: Variation of (t/V) with respect to V

From the equation (7.23), we have

$$k_c = \frac{\mu \alpha W}{A^2 \Delta P} \qquad \text{...(7.23)}$$

$$\Rightarrow \qquad \alpha = \frac{k_c A^2 \Delta P}{\mu W}$$

$$\Rightarrow \quad \alpha = \frac{13076923 \times (0.045)^2 \times 46500}{0.89 \times 10^{-3} \times (13.11)}$$

$$\Rightarrow \quad \alpha = 1.056 \times 10^{11} \text{ m/kg} \quad \textbf{Ans.}$$

From the equation (7.24), we have

$$q_0 = \frac{A\Delta P}{\mu R_m} \qquad \text{...(7.24)}$$

$$\Rightarrow \quad \frac{1}{q_0} = \frac{\mu R_m}{A\Delta P}$$

$$\Rightarrow \quad 27{,}500 = \frac{0.89 \times 10^{-3} \times R_m}{0.045 \times 46500}$$

$$\Rightarrow \quad R_m = 6.465 \times 10^{10} \text{ m}^{-1} \quad \textbf{Ans.}$$

Example 7.2. In a filtration experiment the slurry in water at 25°C is filtered through a plate and frame filter system at a constant pressure of 300 kPa. The filtration time and the obtained filtrate volume are given below:

Sl. No.	V (m^3)	t (s)
1	1.05×10^{-3}	10
2	1.73×10^{-3}	20
3	2.27×10^{-3}	30
4	2.74×10^{-3}	40
5	3.15×10^{-3}	50
6	3.58×10^{-3}	60
7	3.88×10^{-3}	70
8	4.21×10^{-3}	80
9	4.51×10^{-3}	90
10	4.80×10^{-3}	100
11	5.08×10^{-3}	110
12	5.35×10^{-3}	120

The filter area is 0.05 m^2 and the slurry concentration is 20 kg/m^3. Calculate the specific cake resistance 'α' and the filter medium resistance 'R_m'.

Solution: Data:

$$T = 25°\text{C}$$

$$A = 0.05 \text{ m}^2$$

$$\Delta P = 300 \text{ kPa}$$

$$W = 20 \text{ kg/m}^3$$

$$\mu = 0.89 \times 10^{-3} \text{ kg/m.s}$$

Table 7.3: Plate and frame filter press test data

Sl. No.	V (m^3)	t (s)
1	1.05×10^{-3}	10
2	1.73×10^{-3}	20

3	2.27×10^{-3}	30
4	2.74×10^{-3}	40
5	3.15×10^{-3}	50
6	3.58×10^{-3}	60
7	3.88×10^{-3}	70
8	4.21×10^{-3}	80
9	4.51×10^{-3}	90
10	4.80×10^{-3}	100
11	5.08×10^{-3}	110
12	5.35×10^{-3}	120

Using the plate and frame filter press test data given in Table 7.3 perform the calculation and get the corresponding value of t/V as shown in Table 7.4.

Table 7.4: Plate and frame filter press test data

Sl. No.	V (m^3)	t (s)	t/V (s/m^3)
1	1.05×10^{-3}	10	9,524
2	1.73×10^{-3}	20	11,560
3	2.27×10^{-3}	30	13,216
4	2.74×10^{-3}	40	14,598
5	3.15×10^{-3}	50	15,873
6	3.58×10^{-3}	60	16,760
7	3.88×10^{-3}	70	18,041
8	4.21×10^{-3}	80	19,002
9	4.51×10^{-3}	90	19,956
10	4.80×10^{-3}	100	20,883
11	5.08×10^{-3}	110	21,653
12	5.35×10^{-3}	120	22,430

Then, plot the graph t/V vs V as shown in Fig. 7.15.

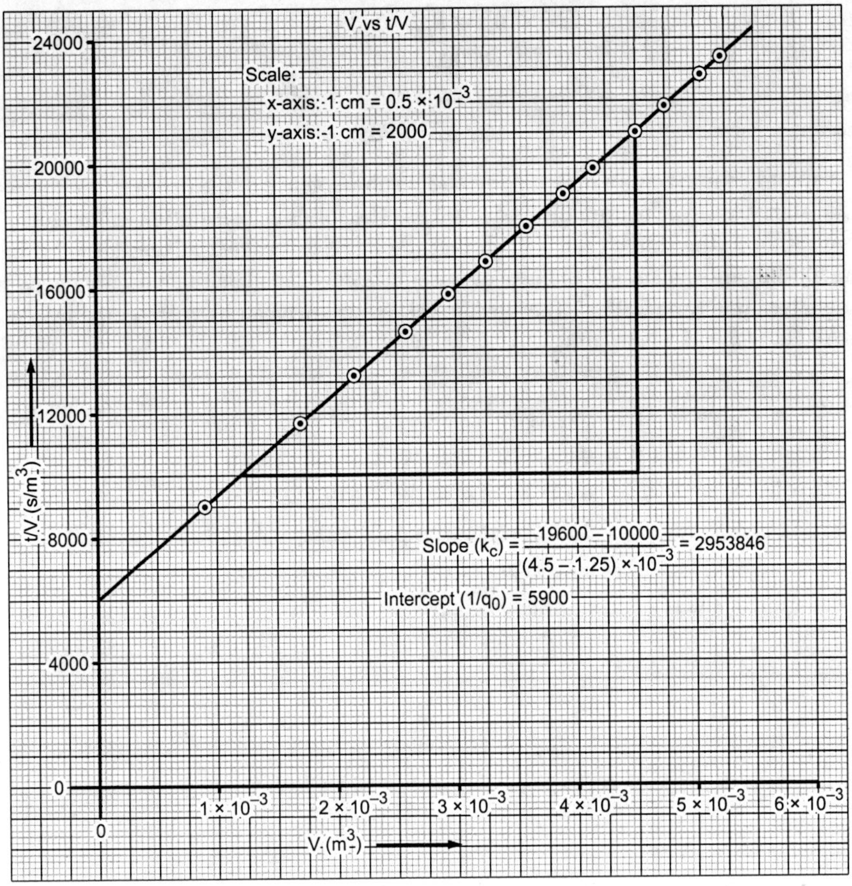

Fig. 7.15: Variation of t/V with respect to V

From the equation (7.23), we have

$$k_c = \frac{\mu \alpha W}{A^2 \Delta P} \quad \text{...(7.23)}$$

$$\Rightarrow \quad \alpha = \frac{k_c A^2 \Delta P}{\mu W}$$

$$\Rightarrow \quad \alpha = \frac{2953846 \times (0.05)^2 \times 300000}{0.89 \times 10^{-3} \times 20}$$

$$\Rightarrow \quad \alpha = 1.245 \times 10^{11} \text{ m/kg}$$

Specific cake resistance, $\alpha = 1.245 \times 10^{11}$ m/kg **Ans.**

From the equation (7.24), we have

$$q_0 = \frac{A \Delta P}{\mu R_m}$$

$$\Rightarrow \quad \frac{1}{q_0} = \frac{\mu R_m}{A \Delta P}$$

$$\Rightarrow \qquad 5900 = \frac{0.89 \times 10^{-3} \times R_m}{0.05 \times 300000}$$

$$\Rightarrow \qquad R_m = 9.943 \times 10^{10}\ \text{m}^{-1}$$

Filter medium resistance, $R_m = 9.943 \times 10^{10}\ \text{m}^{-1}$ **Ans.**

Example 7.3. A rotary drum vacuum filter having a 33% submergence of the drum in the slurry is to be used to filter a $CaCO_3$ slurry using a pressure drop of 67.0 kPa. The solids concentration in the slurry is $C_S = 0.191$ kg solid concentration in the slurry cake is such that the kg of wet cake/kg dry cake = 2. The density and viscosity of the filtrate can be assumed as those of water at 25°C. Calculate the filter area need to filter 0.778 kg slurry/s. The filter cycle time is 250s. The specific cake resistance can be represented by $\alpha = (4.37 \times 10^3)(\Delta P)^{0.3}$, where ΔP is in *Pa* and α is in m/kg.

Solution: For water at 25°C:

$$\rho = 997\ \text{kg/m}^3$$

$$\mu = 0.89 \times 10^{-3}\ \text{kg/m.s}$$

$$C_S = 0.191\ \text{kg of solid/kg slurry}$$

$$m = 2\ \text{kg of wet cake/kg dry cake}$$

$$\Delta P = 67.0\ \text{kPa}$$

From the equation of slurry concentration, we have

$$C_S = \frac{\rho C_s}{1 - mC_s}$$

$$\Rightarrow \qquad C_S = \frac{997 \times 0.191}{1-(2)(0.191)}$$

$$\Rightarrow \qquad C_S = 308\ \text{kg solid/m}^3\ \text{of filtrate}$$

From the given data, we have

$$\alpha = (4.37 \times 10^3)(67 \times 10^3)^{0.3}$$

$$\Rightarrow \qquad \alpha = 1.225 \times 10^{11}\ \text{m/kg}$$

From the flow rate of the filtrate, we have

$$\frac{V}{t} = \frac{0.778 C_s}{C_s}$$

$$\Rightarrow \qquad \frac{V}{t} = \frac{0.778 \times 0.191}{308}$$

$$\Rightarrow \qquad \frac{V}{t} = 4.824 \times 10^{-4}\ \text{m}^3\ \text{filtrate/s}$$

From the rotary drum vacuum filtration, we have

$$\frac{V}{At} = \left[\frac{(m)(\Delta P)}{(t)(\mu)(\alpha)(C_S)}\right]^{1/2}$$

where, V = Volume of the filtrate, m^3

A = Area of the filter, m^2

t = Filtration time, s

m = kg of wet cake/kg of dry cake

μ = Viscosity of the filtrate, kg/m.s

α = Specific cake resistance, m/kg

C_S = kg of solid/m^3 of filtrate

$$\Rightarrow \quad \frac{4.824\times10^{-4}}{A} = \left[\frac{2(0.33)\times(67000)}{250\times(0.89\times10^{-3})\times(1.225\times10^{11})\times(308)}\right]^{1/2}$$

$$\Rightarrow \quad \frac{4.824\times10^{-4}}{A} = \left[\frac{44220}{8.395\times10^{12}}\right]^{1/2}$$

$$\Rightarrow \quad \frac{4.824\times10^{-4}}{A} = 0.0000726$$

$$\Rightarrow \quad A = 6.64 \text{ m}^2$$

Filter area needed to filter is 6.64 m^2. **Ans.**

REVIEW QUESTIONS

7.1 Define filtration.
7.2 Explain the mechanism of filtration.
7.3 What is entanglement? Explain.
7.4 What is impingement? Explain.
7.5 What is straining? Explain.
7.6 Explain the types of filtration.
7.7 Explain briefly the depth filtration.
7.8 Explain briefly the depth filtration.
7.9 What do you mean by cake filtration? Explain.
7.10 Explain theory of filtration.
7.11 Briefly explain the process of constant pressure filtration.
7.12 Briefly explain the process of constant rate filtration.
7.13 Explain the factors influencing the filtration.
7.14 What are the factors which influence pressure drop across the filter medium?
7.15 Briefly explain the term filter media. Give suitable example.
7.16 Briefly explain the term filter aid. Give suitable examples.
7.17 Give some specific application of filtration.
7.18 Classify the filtration equipment.
7.19 With a neat diagram explain the construction and working of sand filter.
7.20 With a neat diagram explain the construction and working of plate and frame filter press.
7.21 With a neat diagram explain the construction and working of leaf filter.
7.22 With a neat diagram explain the construction and working of rotary drum filter.
7.23 With a neat diagram explain the construction and working of horizontal belt filter.
7.24 With a neat diagram explain the construction and working of bag filter.
7.25 With a neat diagram explain the construction and working of meta filter.
7.26 With a neat diagram explain the construction and working of suspended batch centrifuge.
7.27 Differentiate between batch and continuous filter.
7.28 Differentiate between pressure and vacuum filter.
7.29 Explain the characteristics of filter aid.

EXERCISES

7.1 Starting from the basic derive the modified Kozeny Carman equation used for filtration.
7.2 Show that for constant pressure filtration

$$\frac{t}{V} = \frac{k_c V}{2} + \frac{1}{q_0}$$

where, $k_c = \dfrac{\mu \alpha W}{A^2 \Delta P}$

$$q_0 = \frac{A\Delta P}{\mu R_m}$$

7.3 Show that for constant rate of filtration

$$\frac{V}{At} = \frac{A\Delta P}{\mu(\alpha WV)}$$

7.4 In a filtration experiment the slurry in water at 25°C is filtered through a plate and frame filter press system at a constant pressure of 300 kPa. The filtration time and the obtained filtrate volume are recorded as shown in the following table.

Sl. No.	V (m^3)	t (s)
1	1.0×10^{-3}	10
2	1.8×10^{-3}	20
3	2.4×10^{-3}	30
4	2.8×10^{-3}	40
5	3.2×10^{-3}	50
6	3.6×10^{-3}	60
7	3.9×10^{-3}	70
8	4.3×10^{-3}	80

The filter total area is 0.05 m^2 and the slurry concentration is 20 kg/m^3.

Calculate:

(*i*) The specific cake resistance 'α'. (*ii*) Filter medium resistance 'R_m'.

7.5 A rotary drum vacuum filter having a 30% submergence of the drum in the slurry is to be used to filter a $CaCO_3$ slurry using a pressure drop of 70 kPa. The solid concentration in the slurry is C_S = 0.19 kg solid/kg slurry and the filter cake is such that the kg of wet cake/kg dry cake = 2.0. The density and viscosity of the filtrate can be assumed as those of water at 25°C. Calculate the filter area need to filter 0.76 kg slurry/s. The filter cycle time is 250s. The specific cake resistance can be represented by $\alpha = (4.25 \times 10^3)(\Delta P)^{0.3}$, where, ΔP is in *Pa* and α is in m/kg.

7.6 Laboratory filtration conducted at constant pressure drop on a slurry of calcium carbonate in water gave the data as shown below. The filter area was 440 cm^2. The mass of solid per unit volume of filtrate was 23.5 g/l. Evaluate the specific cake resistance.

Sl. No.	Filtrate volume (l)	Time (s)
1	0.5	17.3
2	1.0	41.3
3	1.5	72.0
4	2.0	108.3
5	2.5	152.1
6	3.0	201.7

❑❑❑

8

CHAPTER

Agitation and Mixing

8.1 INTRODUCTION

Success of many processing operations depends on the effective agitation and mixing of their components. Agitation and mixing are not synonymous Agitation refers to the induced motion of a material in a specified way, usually in a circulatory pattern inside the sort of container.

Mixing is the random distribution of the component into and through one another, of two or more separate phases. A single homogeneous such as tank full of cold water can be mixed, but it cannot be mixed until some other material is added to it.

The term mixing is applied to a variety of operations differing widely in the degree of homogeneity of the mixed material. Consider in one case, two gases are brought together to blend thoroughly and in second case sand, gravel, cement and water tumbled in a rotating drum for a long time. In both the cases the final product is said to be mixed. Yet the products are obviously not equally homogenous. Samples of mixed gases, even very small samples all have the same composition. The small samples of the mixed concrete on the other hand differ widely in composition.

8.2 CLASSIFICATION OF MIXING

There are three common classification of mixing. They are as follows:

(*i*) Liquid-liquid mixing
(*ii*) Solid-liquid mixing
(*iii*) Liquid-gas mixing

8.2.1 Liquid-liquid Mixing

It is a process in which mixture is formed from two or more liquid ingredients. Within this category are two additional categories. Miscible and immiscible mixing. Miscible mixing is the complete mixing of two substances to form a single substance.

Example is non-carbonated beverages with flavorings.

Immiscible mixing involves two liquids that do not mix to uniformity. Example is oil and water. When using miscible ingredients in a mixture, the product is considered homogeneous or of uniform structure throughout.

The process for mixing immiscible mixtures have similar goals for their solution. Emulsification often the oil or fatty particles will be reduced to tiny droplets using a high shear mixing blade and will be suspended evenly throughout the mixture, creating a stable mix to eliminate the separation of the two ingradients.

8.2.2 Solid-liquid Mixing

It is a process in which mixing take place between a liquid and solid substances, usually a dissolvable powder or flake, but can also be non-dissolving solids. When a solid substance can be dissolved into a liquid, a solution is created. For example, sweetend milk or sugar water would fall into this category, as they create a homogeneous mixture. This process is referred to as solid suspension and creates a slurry of the suspended solid ingredients.

Waste water processors use this technique to suspend heavy metals in a solution and extract the beneficial ingredients leaving behind slag. The solid suspension process can be used for a variety of reasons but usually to create physical and chemical reactions.

8.2.3 Liquid-gas Mixing

It involves the mixing of suspending gas bubbles thoroughly in an immiscible liquid.

Using a turbine type impeller allows the breaking up of larger bubbles into smaller bubbles to create a larger surface area of a gas and liquid contact. For example, most familiar product like soda or beer, but these solutions can be used in highly technical or volatile applications, across a wide array of industries. Foamy is usually an example of this process like soap or whipped cream.

All application requires a different set of techniques to achieve these different form of mixing. It will be important to adjust shear, agitation level, suspension dispersion and pumping mechanism to best help to achieve an end product.

8.3 APPLICATION OF MIXING

Mixing is one of the most common chemical engineering operations. It is involved in the preparation of different type of formulation. Mixing is also an intermediate stage in the production of several dosage.

(*i*) Wet mixing in the granulation step in the production of capsules.

(*ii*) Dry mixing of several ingredient for direct compression to manufacture tablet.

(*iii*) Dry blending of powder for manufacture of tablets.

(*iv*) Production of pellets for capsules.

In the manufacture of tablets, normally a number of additives are added. Therefore, mixing of powders becomes an essential part of the process. When the dose of the active substance is high mixing is not a problem. But, in case of potent drugs high amount of adjuvants are added. Therefore, mixing is considered as a critical factors. Otherwise, content uniformity does not confirm to the pharmaceutical specifications. Similarly, weight variation increases.

8.4 FACTORS INFLUENCING MIXING

Particles characteristics influence the mixing process. Aggregation inhibits proper mixing. Therefore, higher shearing forces are applied. Hence, correct mixing of one component does not imply good mixing of other components. Therefore, adding dye to the mixture is often misleading. The dye may not aggregate, but active substance may aggregate. A single factor can in no way be considered as a unique indication of mixing. However, flow properties of the components is the most important consideration which is again influenced by a number of factors:

(*i*) Nature of the surface

(*ii*) Density of the particles

(*iii*) Particle size

(*iv*) Particle shape

(*v*) Particle charge

(*vi*) Proportion of the materials

8.4.1 Nature of the Surface

Rough surface of one of the components does not induce satisfactory mixing. This can be due to the entry of active substance into the pores of the other ingredients. Adding a substance, which will be adsorbed on its surface, can decrease aggregation.

For example, addition of aerosol to zinc oxide. Thus, a strongly aggregating zinc oxide becomes a fine dusting powder which can be mixed easily.

8.4.2 Density of the Particles

The density of the particle has minor importance. Demixing is accelerated when the density of the smaller particle is higher or when the mixing process is stopped abruptly. This is due to the fact that dense material always moves downward and settle at the bottom.

8.4.3 Particle Size

It is easy to mix two components having approximately the same particle size. The variation of particle size can lead to separation, because the small particles move downward through the space between the bigger particles. As the particle size increases, flow properties also increases due to the influence of gravitational force on the size. Beyond a particular point, flow property decreases. The component with a mean particle size less than 10 μm are free flowing which facilitates mixing.

8.4.4 Particle Shape

The ideal particle is spherical in shape for the purpose of uniform mixing. The irregular shapes can become interlocked and there are less chances of separation of particles once these are mixed together.

8.4.5 Particle Charge

Some particles exert attractive forces due to electrostatic charges on them. This can lead to separation or segregation.

8.4.6 Properties of Materials

The best result in mixing can be obtained when two components are to be mixed in equal proportion by weight and by volume. If there is a large difference in the proportion of the two component, then mixing is always done in the ascending order of their weights.

8.5 MIXING EQUIPMENT

The equipment used for agitation and mixing are as follows:

(*i*) Agitated tank

(*ii*) Airjet mixer

(*iii*) Jet mixer

(*iv*) Flow mixer

(*v*) Tumbler blender

(*vi*) Double cone tumbling blender
(*vii*) Ribbon blender
(*viii*) Internal screw mixer
(*ix*) Muller mixer
(*x*) Planetary mixer
(*xi*) Kneader mixer
(*xii*) Banbury mixer
(*xiii*) Barrel mixer
(*xiv*) Zigzag blender
(*xv*) Silverson mixer emulsifier
(*xvi*) Ultrasonic emulsifier
(*xvii*) Gas-liquid mixer

8.5.1 Agitated Tank

Liquids are most often agitated in some kind of tank, usually cylindrical in form with a vertical axis. A typical agitated tank is shown in Fig. 8.1.

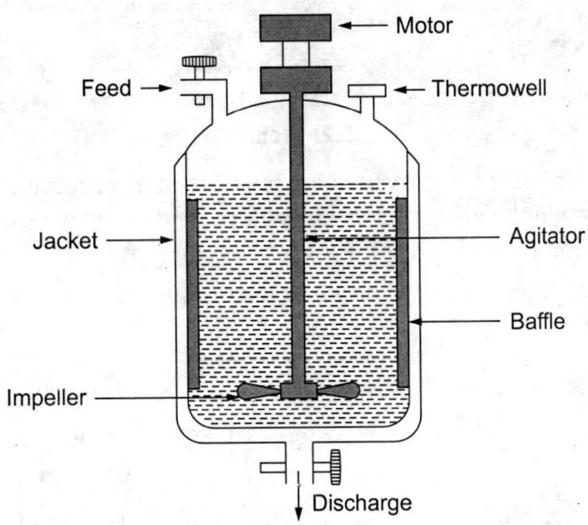

Fig. 8.1: Agitated tank

The top of the vessel may be open, usually it is closed. The proportion of the tank varies widely depending on the nature of the agitation. The tank bottom is rounded not flat to eliminate sharp corner or regions into which fluid current would not penetrate. The liquid depth is approximately equal to diameter of the tank. An impeller is mounted on an overhung shaft that is supported from the top of the tank. The shaft is driven by motor sometime directly connected to the shaft but more often connected to it through a speed reducing gear box. Accessories such as inlet and outlet lines, coils jacket and well for thermometer or other temperature measuring devices are usually included.

The impeller causes the liquid to circulate through the vessel and eventually returns to the impellers. Baffles are often included to reduce tangential motion.

Impellers: There are two type of impellers

(*i*) Axial flow impeller
(*ii*) Radial flow impeller

Axial flow impeller generate flow currents parallel to the axis of shaft and radial flow impellers are those which generate flow currents in tangential or radial direction.

The main types of impellers are:

(*i*) Propellers

(*ii*) Paddles

(*iii*) Turbines

Propeller and pitched blade turbines are axial flow impellers while the paddle and flat blade, disk flat blade turbines are radial flow impellers.

Propellers: A propeller is an axial flow, high speed impeller commonly used for low viscosity liquids. It may be mounted centrally, off centre or at an angle to the tank. It is simple and portable. The diameter of propeller is usually between 15 to 30% of the diameter of tank. A typical propeller is shown in Fig. 8.2.

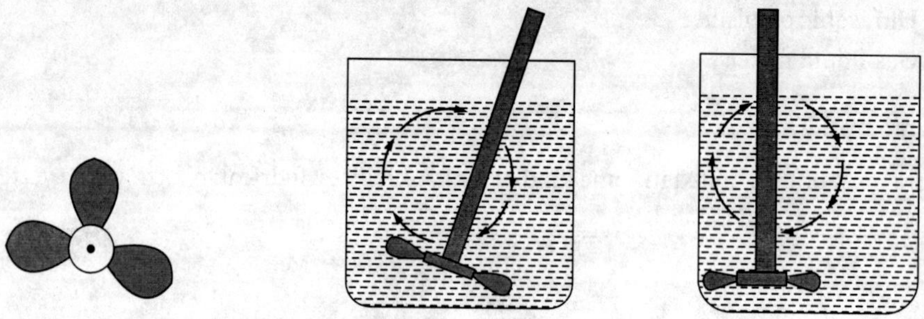

(*a*) Standard three blade propeller (*b*) Angular Propeller (*c*) Off centre propeller

Fig. 8.2: Propellers

Standard three blade marine propellers with square pitch are in most common use. A propeller is shaped with a tapering blade to maintain the effect of centrifugal force and produce maximum axial flow. The small propellers are rotates at full motor speed, while the bigger one rotate at speed of 400 to 800 RPM.

Propellers may also be mounted near the bottom of the cylindrical wall of the vessel. A typical propeller of this type is shown in Fig. 8.3.

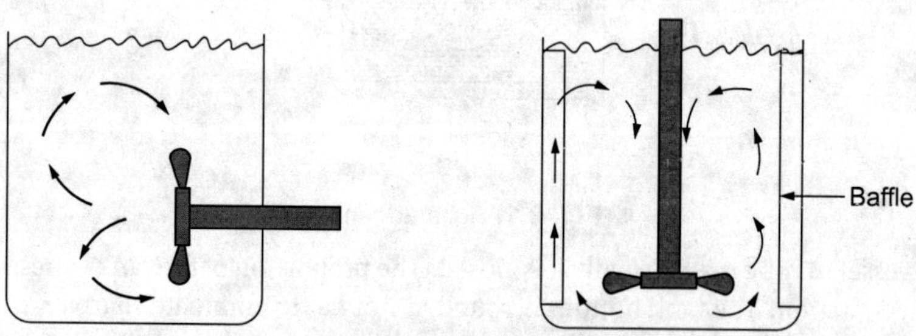

(*a*) Side entering propeller (*b*) Baffled vessel with centrally mounted propeller

Fig. 8.3: Propellers

It is used for blending low viscosity fluids or suspending a slow settling sediments in very large tank. Propellers drives the liquid straight down to the bottom of the tank, at the bottom the stream spreads radially in all direction towards the wall, then the liquid flows upwards along the wall and finally returns to the suction of impeller from the shaft. This type of propellers are used in a situations where strong vertical currents are desired for example suspending heavy particles.

Paddles: For a simple situation a flat paddle turning on a vertical shaft is used. Paddle agitator with two or four blades are more common. The blades of these agitators normally extends close to the tank wall. They are simply pusher and cause the mass to rotate in laminar swirling motion with practically no radial flow along the paddle blades or any axial flow. The circulation is poor and the mixing action is insufficient. The speed of rotation is very low and is generally between 30 to 150 RPM. The total length of the paddle impeller is typically 50 to 80% of the inside diameter of the vessel. The width of the paddle is 1/4 to $1/10^{th}$ of the paddle diameter.

In some case, the shape of the blade is more or less same as the shape of a dished or hemispherical vessel so that they scrap the surface or pass over it with the close clearance. A paddle of this kind is known as anchor agitator. A typical paddle agitator is shown in Fig. 8.4.

(*a*) Flat paddle (*b*) Anchor

Fig. 8.4: Paddle agitator

Anchors are very useful for preventing deposits on heat transfer surface as in reaction vessels and are commonly employed for obtaining improved heat transfer in high viscosity fluid. High viscosity liquids and pastes are agitated by multiple paddle blades. Paddles cover wide range of viscosity, not easily destroyed in operation and are relatively low in cost.

Turbines: Various types of turbine impellers are available. Among these pitched blade turbine is an axial flow impeller while curved blade and flat blade turbines are radial flow impellers. They are capable of creating a vigorous mixing action due to centrifugal and rotating motion generated by them. A stator ring surrounding this impeller gives an efficient mixing action. A typical turbine impeller is shown in Fig. 8.5.

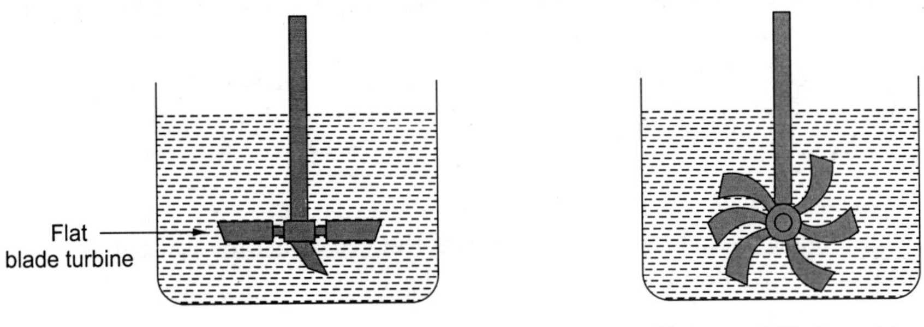

(*a*) Pitched blade turbine (*b*) Curved blade turbine

Fig. 8.5: Turbine propeller

The blades of the impeller may be attached to the central hub or to central disc. The diameter of the impeller is kept between 1/3 to $1/6^{th}$ of the vessel diameter. The blade length is one $1/4^{th}$ of the impeller diameter. With central disc, it is $1/8^{th}$ of the impeller diameter. The blade angle of curved blade turbine may be between 30 to 60°. The impeller speed usually ranges from 50 to 250 RPM.

Swirling and Vortex formation: When a low viscosity liquid is stirred in an un baffled tank by centrally mounted agitator, there is a tendency for a swirling to develop for lighter fluid to be drawn in to form a vortex at the top surface of the liquid. This will affect the degree of agitation and hence mixing reduced. This phenomena take place in unbaffled tanks regardless of type of impeller used. A typical flow pattern in unbaffled tank for either axial or radial flow impeller is shown in Fig. 8.6.

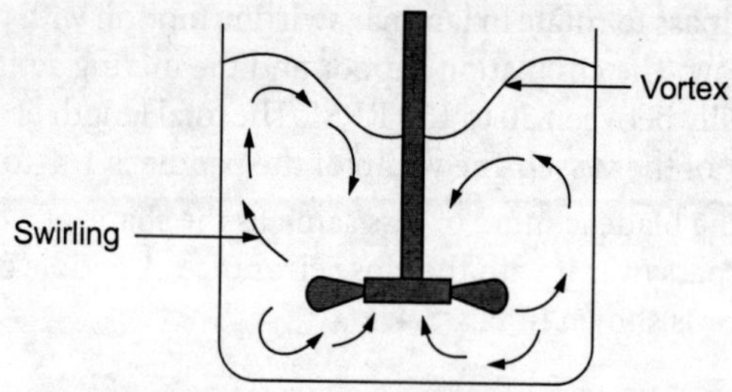

Fig. 8.6: Circulation pattern and vortex formation in an agitated tank.

In a vortexing low viscosity liquid, vertical velocities are low relative to circumferential velocity in the vessel. In vortexing the top surface of the liquid takes roughly U-shape and efficient mixing do not takes place. A vortex is produced owing to centrifugal force acting on a rotating liquid. However, there is a limit to the rotational speed that may be used, since once the vortex reaches impeller, severe air entrainment may occurs. A typical vortex formation at very high impeller speed is shown in Fig. 8.7.

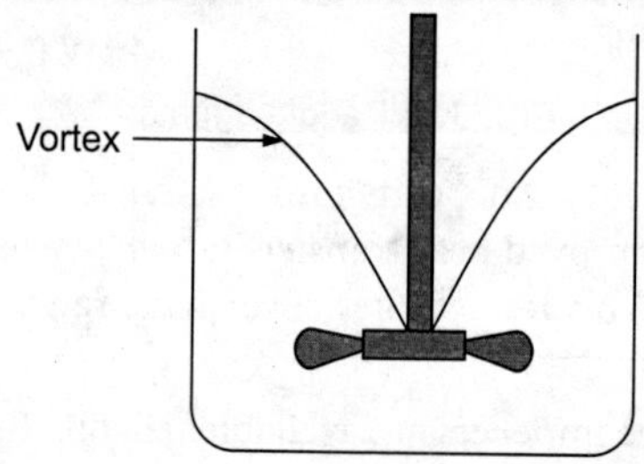

Fig. 8.7: Vortex at very high impeller speed

In addition, when the swirling mass of the liquid generates an oscillating surge in the tank which fluctuating force acting on a agitator shaft. Therefore, for better mixing it is necessary to prevent swirling and vortex formation.

There are three methods of prevention of swirling and vortex formation. They are:

(*i*) Off centre mounting of the impeller

(*ii*) Use of baffles

(*iii*) Use of diffuses ring with turbine

In small tank, the impeller can be mounted off centre. This will reduce the swirling and vortex formation. A typical off centre mounting of the impeller is shown in Fig. 8.8

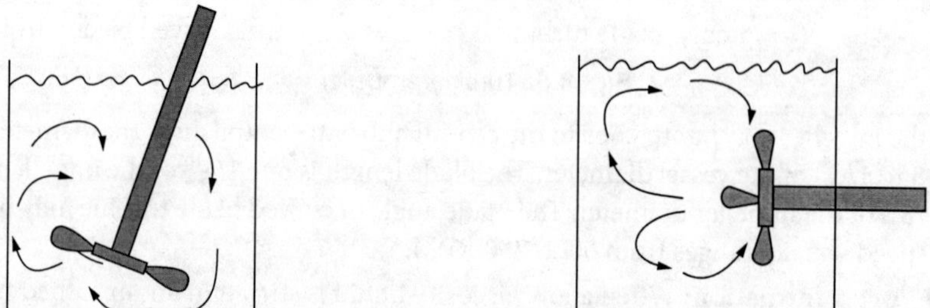

(*a*) Off centre angular propeller (*b*) Side entering propeller

Fig. 8.8: Off centre mounting impeller

In large tank with vertical agitator, most common method of reducing swirling is to install baffle which impede rotational flow without interfering with radial or longitudinal flow. A typical baffled tank with centrally mounted impeller is shown in Fig. 8.9.

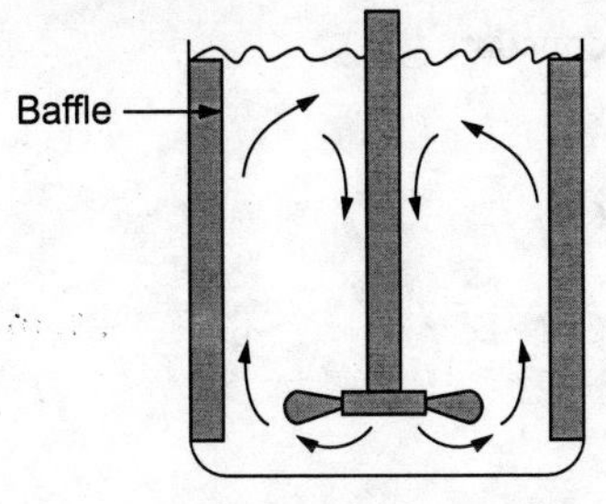

Fig. 8.9: Baffled tank with centrally mounted impeller

In unbaffled vessel, there are strong tangential flow and vortex formation is at moderate speed. In the presence of baffles the vertical flows increased and hence more rapid mixing of the liquid.

In case of turbine, the currents are radial and tangential. The tangential components induce swirling and vortexing that must be stopped by use of diffuser ring for impeller to be most effective.

Power consumption of impellers: The empirical correlation of power with other variable of the system permit fairly accurate prediction of power requirement of an impeller to rotate at a given speed. Such correlations can be obtained by using a method of dimensional analysis.

Power requirement of the impeller is a function of geometrical details of the impeller and vessel, viscosity and density of the fluid and rotation speed of the impeller. The empirical correlation that can be obtained for power consumption of impeller by dimensional analysis may be expressed as:

$$\frac{P}{N^3D^5\rho} = \phi\left[\frac{ND^2\rho}{\mu}, \frac{N^2D}{\rho}\right] \quad ...(8.1)$$

where, $\frac{P}{N^3D^5\rho}$ = Power number, N_P

$\frac{ND^2\rho}{\mu}$ = Impeller Reynolds number, N_{Re}

$\frac{N^2D}{\rho}$ = Froude number, N_{Fr}

N = Rotation speed in revolution

D = Diameter of the impeller, m

ρ = Density of the fluid, kg/m^3

μ = Viscosity of the fluid, kg/m.s

At lower Reynolds number, the relationship between N_P and N_{Re} may be expressed as

$$N_P = \frac{C_o}{N_{Re}} \quad ...(8.2)$$

$$C_o = N_P \cdot N_{Re} \quad ...(8.3)$$

Substituting the values of the N_P and N_{Re} in equation (8.3), we get

$$C_o = \frac{P}{N^3D^5\rho} \times \frac{ND^2\rho}{\mu}$$

$$\Rightarrow \quad P = C_o N^2 D^3 \mu \quad \text{...(8.4)}$$

Equation (8.4) indicates that if the speed is doubled, power consumption will increase by a factor of four.

For higher values of N_{Re}, the Froude number plays an important role. In this, the power number is constant,

$$N_P = \text{constant}$$

$$N_P = C$$

$$\Rightarrow \quad \frac{P}{N^3 D^5 \rho} = C$$

$$\Rightarrow \quad P = C\rho N^3 D^5 \quad \text{...(8.5)}$$

8.5.2 Airjet Mixer

When compressed airjet is passed from the bottom of the vessel, air bubbles are formed in the liquid phase. The buoyancy of the bubbles lift the liquids, which is confined to the central part of the tank due to the presence of draft tube. The liquid flows down from the periphery of the vessel and enter from the bottom due to suction effect. A typical airjet mixer is shown in Fig. 8.10.

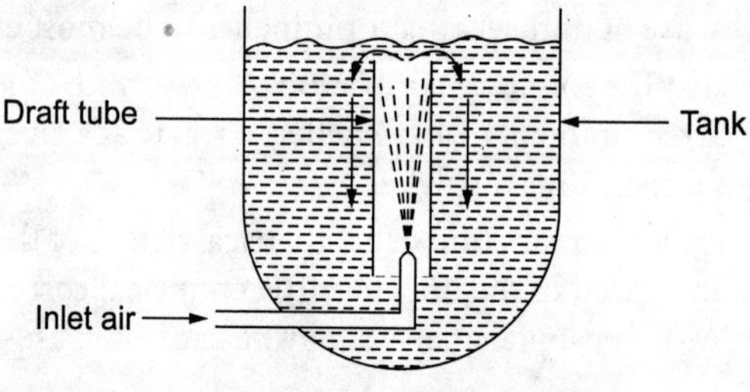

Fig. 8.10: Airjet mixer

The liquid is placed in the vessel. Draft tube is placed surrounding airjet. Compressed air is allowed to pass at high pressure from the inlet provided at the bottom of the tank. This causes the buoyancy of the bubbles, which lift liquid from the bottom of the top of the vessel.

Draft tubes serve to confine the expanding bubbles and entrained liquid to the central part of the vessel. This results in a more efficient lifting action by the bubbles. The liquid mixture flown down from the periphery of the vessel.

The overall circulation brings liquid from all parts of the vessel of the region of the jet itself. Thus mixing is achieved.

The liquids of low viscosity, non-foaming, non-reactive with gas are mixed by using airjet mixer.

8.5.3 Jet Mixer

The liquid to be mixed are pumped separately into a jet mixer at different velocities. The high velocity fluid has a lower static pressure than surrounding liquid. During mixing, the slow moving liquid will be drawn into the high velocity jet. This increases the volume of the jet and decreases velocity, which causes the jet to expand. A typical jet mixer is shown in Fig. 8.11.

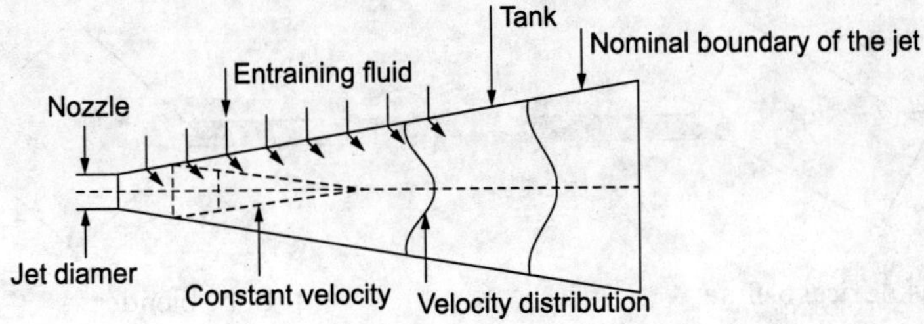

Fig. 8.11: Jet mixer

Two liquids are introduced into the mixing tank at different velocities. One liquid is pumped through a small nozzle at uniform high velocity. The second liquid enters the tank at a low velocity. The fast moving liquid impinges on the slow moving liquid at high velocity. As the jet moves away from the surface, the area of its influence decreases. The core of the jet is surrounded by an expanding turbulent jet in which the radial velocity decreases with distance from the centre line of the jet.

When a rapidly moving liquid comes in contact with slow moving liquid, a velocity gradient developed at the boundary. As a result, strong shear stresses developed at the boundary which induce mixing.

Enough time and space required for the stream to be blended thoroughly into the mass of fluid by the mechanism of entrainment.

8.5.4 Flow Mixer

It consists of a pipe in which baffles are attached. A typical flow mixer is shown in Fig. 8.12.

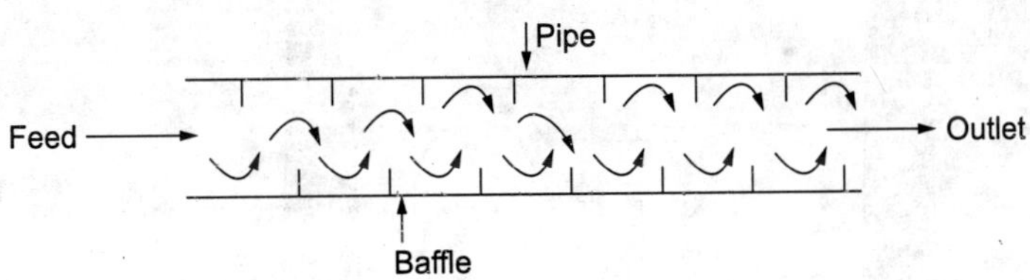

Fig. 8.12: Flow mixer

Liquids to be mixed are passed through the pipe. Mixing takes place mainly through bulk transport in the direction of flow. Placing of baffle, enhances the mixing efficiency.

For an effective mixing, controlling the feed rate is essential. For this purpose suitable metering devices are employed. If input rate is difficult to control and fluctuations in the added proportions of the liquids are unavoidable, continuous mixing equipment of tank type is preferable, because hold up of the liquids and back flow or recirculation is possible.

8.5.5 Tumbler Blender

It is a general type of equipment meant for blending of dry solids. The equipment consists of a container of one of the several geometric form. These are mounted so that it can be rotated about an axis. These do not have packing glands round the shaft. The cubes or hexagonal cylinder may be rotated in any axis depending on the manufacture. A typical tumbler blender is shown in Fig. 8.13.

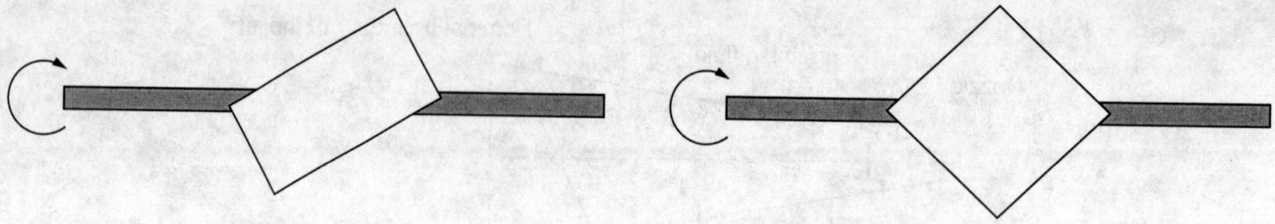

(*a*) Cylindrical blender (*b*) Cube blender

Fig. 8.13: Tumbler blender

Such a cylindrical vessel is not suitable for mixing because the movement of the particle is in one dimensional. Neither cylindrical nor cube mixer are used to a some extent in the industry.

8.5.6 Double Cone Tumbling Blender

Many material are mixed by tumbling in a partly filled container, which rotates about it horizontal axis. A typical double cone tumbling blender is shown in Fig. 8.14.

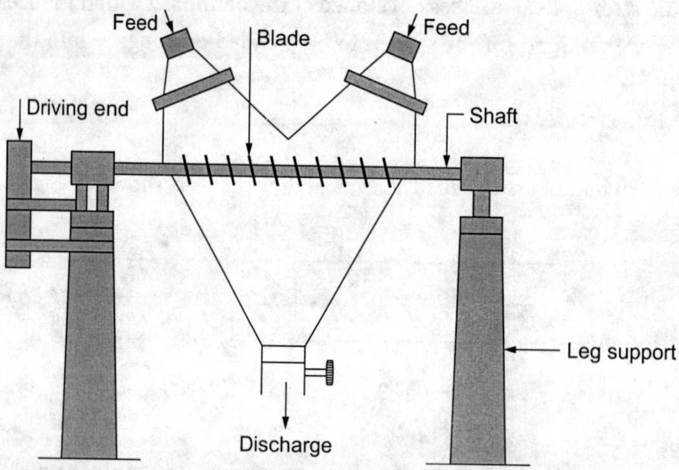

Fig. 8.14: Double cone tumbling blender

It is also called twin shell tumbling blender. The double cone tumbling blender is suitable for free flowing dry powder mixing.

It consists of two cone, base to base with or without cylindrical section in between. The mixer is mounted so that it can be rotated about an axis perpendicular to the line joining the points of the cones.

The material to be mixed is charged to the mixer from the feed inlet until it is 50 to 60% full. The top of the container is closed and the solids are tumbled for a period of about 5-20 minutes. Finally, the mixed material is dropped out from the bottom of the container into a conveyor bin.

The twin shell blender is made from two short cylinder joined so that their axes are about 90° to each other and rotate about a horizontal axis. It may have internal spray for introducing small amounts of liquid into the mix for breaking up agglomerates. Tumbling mixer is capable of handling large volume of material to be mixed.

8.5.7 Ribbon Blender

It consists of non-movable horizontal cylindrical trough usually open at the top. The blender is fitted with two helical blades which are mounted on the same shaft through the long axis of the trough. The blades are connected to a fixed speed drive. The ribbon blender is top loading with a bottom discharge port. The trough can be closed with a lid. A typical ribbon blender is shown in Fig. 8.15.

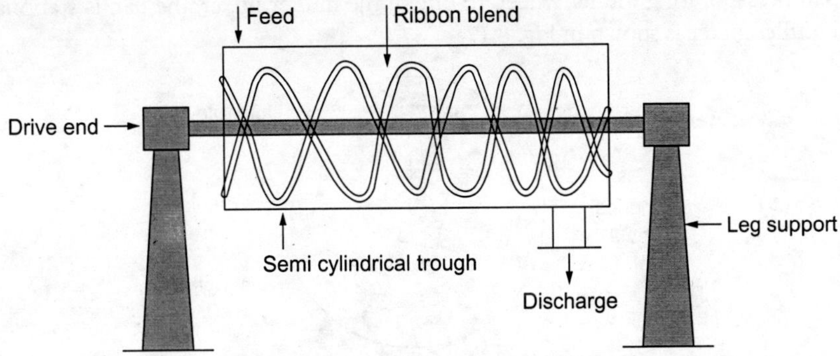

Fig. 8.15: Ribbon blender

Through the fixed speed drive, ribbons are allowed to rotate. One blade move the solids slowly in one direction, while the other one moves solids in other direction. The ribbon may be continuous or interrupted. Mixing takes place due to the turbulence induced by counteracting ribbons, not by mere motion of the solids through the trough.

The ribbon blender may be used for batch or continuous mixing. In batch operated ribbon blenders, the solids are charged and mix until satisfactory and discharged from the bottom. In continuously operated unit, the solids are fed from one end of the trough and discharged from the other end. In the path from feed to discharge end solids are mixed.

For light duty, the trough is open or lightly covered, while the operation under pressure or vacuum the trough is closed. Ribbon blenders are very effective for thin pastes and for powders that do not flow readily.

8.5.8 Internal Screw Mixer

The screw mixer has several variations. A typical type of internal screw mixer is shown in Fig. 8.16.

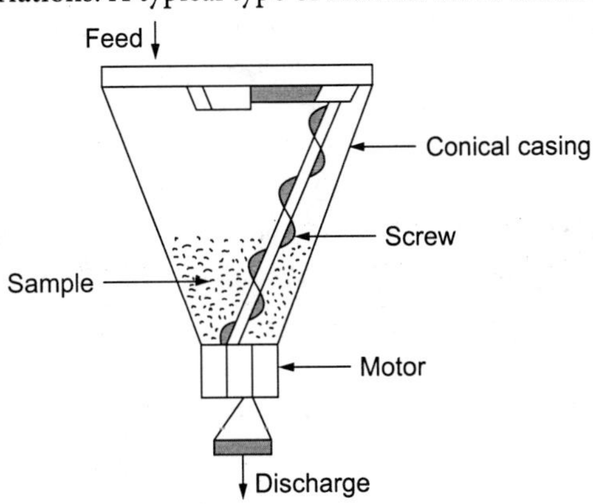

Fig. 8.16: Internal Screw mixer

In this type the screw rotates about its own axis, while orbiting around the centre axis of the central tank.

In another variation, the screw does not orbit but remain in the centre of the conical tank and is tapered so that the swept area steadily increases with increasing height.

In another type the central screw is contained in a inner cylindrical casing.

This type of mixer is promptly suitable for free flowing dry solids.

8.5.9 Muller Mixer

It consists of a pan incorporating muller wheel. In one of the muller mixer, the pan is stationary and wheel rotate. A typical muller mixer is shown in Fig. 8.17.

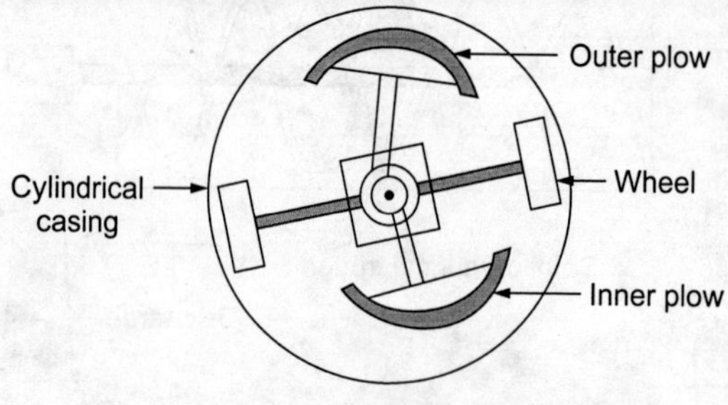

Fig. 8.17: Muller mixer

In other design pan is rotated and axis of the wheel is held stationary. In stationary pan muller mixer, central vertical shaft is driven causing the muller wheels to roll in a circular path over a layer of solids on the pan floor. Plows guide the solids under the muller wheels during mixing or to an opening in the pan floor for discharge of the mixer at the end of the cycle. The muller wheel crush the material, breaking down lumps and agglomerates.

The capacity of the muller mixer ranges from a fraction of cubic meter to more than 1.5 m^3 and the corresponding power requirement ranges from 10 to 75 hp. Muller is a good mixer for handling batches of heavy solids and pastes and especially effective in uniformly coating the particles of granular solids with small amount of liquid.

8.5.10 Planetary Mixer

In a planetary mixer, the blade tears the mass apart and shear is applied between a moving blade and a stationary wall. The mixing arm moves in two ways around its own axis and around the central axis, so that it reaches every spot of the vessel. The plates in the blade are sloped so that the powder makes an upward movement. Therefore, tumbling motion is also obtained.

It consists of a vertical cylindrical shell which can be removed either by lowering it beneath the blade or raising the blade above the bowl. The mixing blade is mounted from the top of the bowl. The shaft is driven by a planetary gear. A typical planetary mixer is shown in Fig. 8.18.

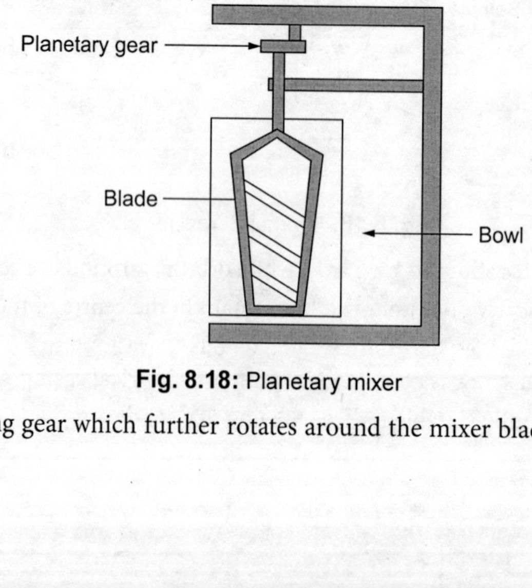

Fig. 8.18: Planetary mixer

It rotates around the ring gear which further rotates around the mixer blade. It is normally built with a variable speed drive.

In planetary mixer the agitator has a planetary motion. It rotates on its own and around the central axis so that it reaches all parts of the vessel. Blade is shaped to pass with close clearance over the side and bottom of the mixing bowl. Therefore, there are no dead spaces in the mixing bowl. The blade tears the mass apart and shear is applied between the moving blade and the stationary wall. The plates in the blades are sloped so that the material makes as upward movement. Therefore, tumbling motion is obtained. Since it is a variable speed driven, initially the blade moves slowly for premixing and finally at increased speed for active mixing. Thus high shear can be applied for mixing.

Emptying the bowl may be done by manually or by dumping mechanism.

Planetary mixer produces precise blend in addition to breaking down of agglomerates rapidly. Low speeds are used for dry mixing and kneading action is required for wet granulation. Steam jacketed bowls are used in the manufacture of sustained release products and ointments. One main dis-advantage of the planetary mixer is that it generate heat within the product.

8.5.11 Kneader Mixer

It consists of a rectangular trough curved at bottom to form two longitudinal half cylinder and a saddle section. Within this trough counter rotating blade roughly Z-shaped outline are incorporated. The blades are so placed that the material turned up by one blade is immediately turned under the adjacent one. The blades are driven by gearing at either one or both the ends. It can be operated batch wise and may be designed to consume very large amount of power. The trough may be open or closed and may be jacketed for heating or cooling.

The mixer can be emptied through the bottom valve where 100% discharge take place. Because of this reason through cleaning between the batch is not essential. In general the double arm Kneader discharge by power operated jacks. A typical double arm Kneader mixer is shown in Fig. 8.19.

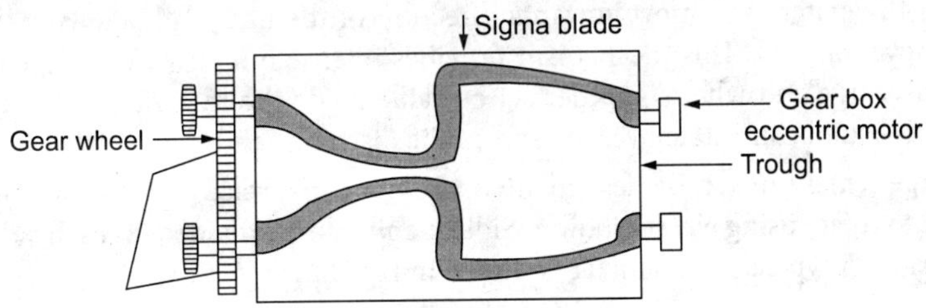

Fig. 8.19: Kneader mixer

The material to be kneaded is dropped into the trough and mixed the content for a period of about 5 to 20 minute or longer. The trough is then unloaded by tilting it. In kneading mixer, the mixing action is combination of bulk movement, smearing, stretching, folding, dividing and recombining as the material is pulled and squeezed against the blades, saddle and the walls of the trough.

Various design of mixing blade are most widely used. The mixer employing sigma blade is capable of starting and operating with either liquid or solid or combination of both. The sigma blade has good mixing action and is relatively easy to clean when sticky materials are being handled.

8.5.12 Banbury Mixer

It is a internal mixer. It is a heavy duty machine with two blades each rotating in cylindrical chamber, but these cylinders partly intersect with each other. In this mixer, the blade is pear shaped, but the projection is spiral along the axis and the two spirals interlock. The mixer operates at a speed of 40 RPM or lower. The clearance between the blades and the walls is extremely small so that the mixing action can takes place. A typical Banbury mixer is shown in Fig. 8.20.

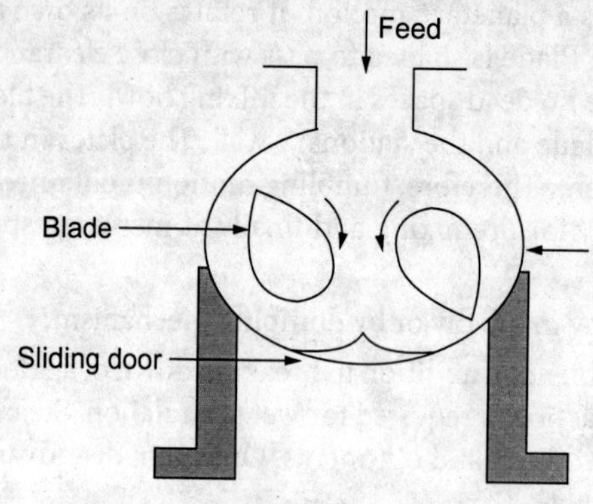

Fig. 8.20: Banbury mixer

The material is fed from the top and held in trough during mixing by an air operated piston under a pressure of 1 to 10 kg/cm^2. The mixed material is discharged through a heavy sliding door which is provided at the bottom of the trough.

The heat generated is taken out by operating cooling water on the wall of the mixing chamber and circulating through the hollow agitator shaft during the operation.

The Banbury mixer is used mainly in plastic and rubber industries.

8.5.13 Barrel Mixer

It is a continuous mixer. The rotating shell keeps the material under tumbling motion. The presence of baffle further enhances the mixing action. When the material approach the mid-point of the shell, another set of baffle causes part of the material to move towards direction of the inlet end, allowing the remaining part of the material to move forward. This mechanism provides intense mixing of the ingredients. The process is continued upto discharge end where, another set of baffle guide the material to the discharge port. The blended material at an equilibrium state overflows from the discharge end.

It resemble a large cement mixer. Baffles are fitted to the inner surface of the shell. The shell is fixed to a shaft, which allowed to rotate using electric power. Side openings are provided on each side for charging and discharging of materials. A typical Barrel mixer is shown in Fig. 8.21.

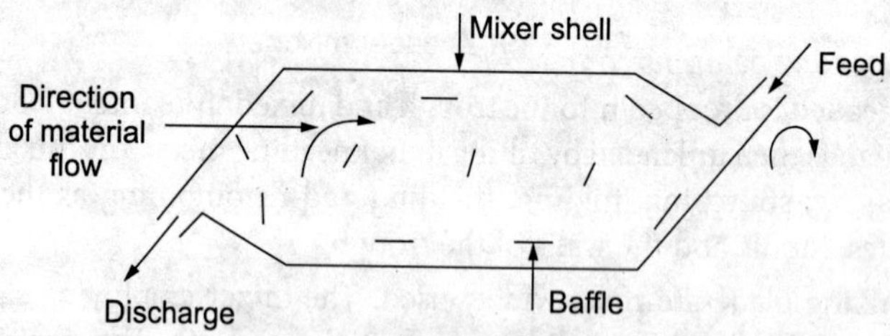

Fig. 8.21: Barrel mixer

The mixer is allowed to rotate. The baffles are also rotate inside the shell. The ingredients are introduced into the shell through the inlet. They begin moving towards the opposite end of the blender due to tumbling action. Baffles further enhance the mixing. As the material approaches the mid-point of the mix shell, the baffles are so positioned to cause the feedback of part of the material in direction of the inlet end. The feedback action continues upto the discharge end. The overflow of material causes the blended solids to discharge.

8.5.14 Zigzag Blender

It is a continuous blender. It consists of a rotating shell with a several *V*-shape blender connected in series. The material undergoes tumbling motion. When the *V*-shape is inverted, the material splits into two portions, one half of the material moves backward, while the other half moves forward. Due to inclined axis of the shell towards the discharged end, the material gets discharged. As the first V-section clears the charge, a fresh feed enters. A typical zig-zag blender is shown in Fig. 8.22.

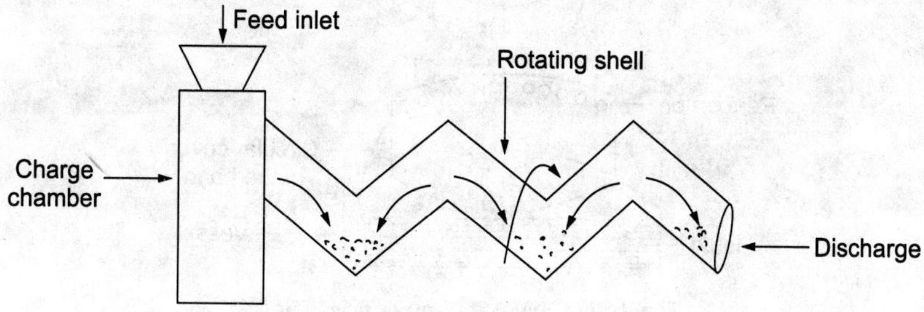

Fig. 8.22: Zig-zag blender

The shell takes the shape of several *V*-shaped blender connected in series. A chamber for charging is attached to one end of the shell. The other end allows the discharge of material. The shell is inclined towards the discharge end.

A single pre-weighed charge is introduced into the charge chamber. As the shell rotates, at one particular angle a metered material enters the *V*-shaped section by gravity. With rotation, one-half of the blend material moves downwards back to the preceding leg and another half of the material moves towards the discharge end. As the first charge clears the first *V*-section, which may take only a few minutes, the next charge is added. The uniformity of blending depends on the flow properties of the ingredients.

8.5.15 Silverson Mixer Emulsifier

Silverson mixer emulsifier produces intense shearing forces and turbulence by the use of high speed motor. The turbulence causes the liquid to pass through fine interstices formed by closely placed perforated metal sheets. Circulation of material takes place through the head by the suction produced in the inlet at the bottom of the head. Circulation of the material ensure rapid breakdown of the dispersed liquid into smaller globules.

It consists of long supporting columns connected to a motor which gives support to the head. The central portion contains a shaft one end of which is connected to the motor and the other end is connected to the head. The head carries turbine blades. The blades are surrounded by a mesh, which is further enclosed by a cover having openings. A typical Silverson mixer emulsifier is shown in Fig. 8.23.

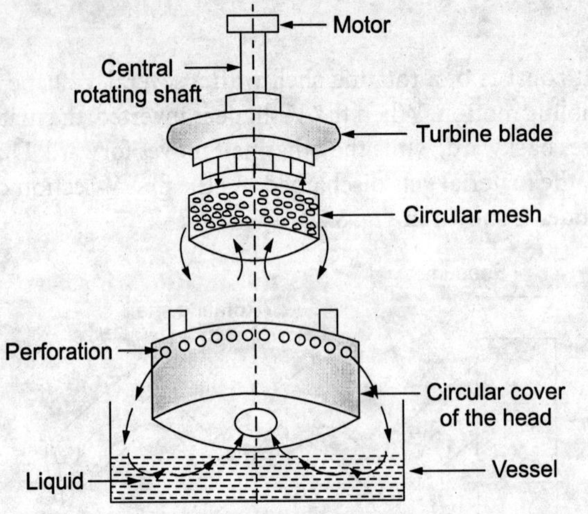

Fig. 8.23: Silverson mixer emulsifier

The emulsifier head is placed in the vessel containing immiscible liquids in such a way that it should be completely dipped in the liquid. When the motor is started, the central rotating shaft rotates the head, which in term rotates the turbine blades at a very high speed. This create a pressure difference. As a result, liquids are sucked into the head from the centre of the base and subjected to intense mixing section. Centrifugal force expel the contents of the head with high force through the mesh and on to the cover. As a result a fine emulsion emerges through the openings of the outer cover. The intake and expulsion of the mixture set up a pattern of circulation to ensure rapid breakdown of the bigger globules into smaller globules.

8.5.16 Ultrasonic Emulsifier

When a liquid is subjected to ultrasonic vibration, alternate regions of compression and rarefaction are produced in the liquid. Cavities are formed in the regions of rarefaction, which subsequently collapse in the region of compression. This results in the generation of high force for emulsification.

It consists of a pump driven by a motor. It is connected with an inlet tube and outlet tube for the discharge of a fine emulsion. The homogenizer head consists of a flat jet for liquid inlet. Facing the jet, a thin blade is preset with edges facing each other. This blade vibrates at its natural frequency of about 30 kilohertz. A typical ultrasonic emulsifier is shown in Fig. 8.24.

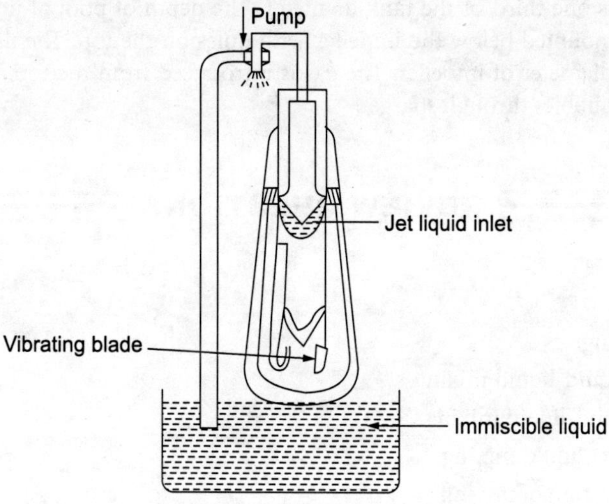

Fig. 8.24 Ultrasonic emulsifier

Coarse emulsion is pumped into Rapisonic homogenizer through one end of the tube. A powerful stream of liquid is forced through the jet. Liquid impinges on the blade causing it to vibrate. The streaming liquid deflects on either side alternatively with the result that oscillation above the sonic range are produced in the liquid. During this process, the emulsion experience alternate region of components and rarefaction. In the region of rarefaction, cavities are formed. Because of external pressure, the cavities collapse violently afterwards. Thus sufficient turbulence is created which is capable of causing dispersion of phases. Thus a coarse emulsion is converted into a fine emulsion.

8.5.17 Gas Liquid Mixer

This usually accomplished by spraying the gas under the turbine near the bottom of the cylindrical vessel. Injecting the gas under a propeller is useless because the flow from a propeller is axial and downward. The equipment which can be used for mixing of gases with liquid consists of baffled vertical vessel incorporating flat blade turbine agitator. A typical agitated vessel for gas-liquid mixing is shown in Fig. 8.25.

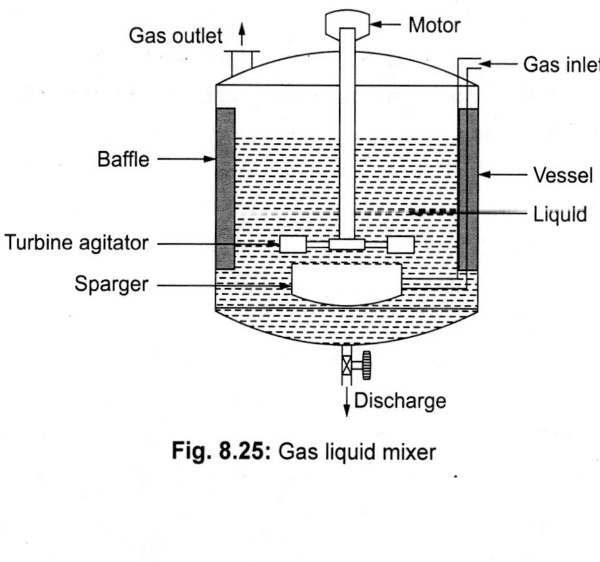

Fig. 8.25: Gas liquid mixer

The diameter of turbine is one third of the tank diameter. The depth of pool of liquid is equal to the tank diameter. The sparger is mounted below the impeller with holes on the top. The diameter of the sparger is equal to or less than the diameter of impeller. The gas is introduced from the top and injected in a pool of liquid in the form of fine bubbles through sparger.

REVIEW QUESTIONS

8.1 Define mixing.
8.2 Define agitation.
8.3 Classify the mixing.
8.4 Briefly explain liquid-liquid mixing.
8.5 Briefly explain solid-liquid mixing.
8.6 Briefly explain gas-liquid mixing.
8.7 Discuss on application of mixing.
8.8 What are the factors influencing mixing? Explain.
8.9 Name the all types of mixing equipment.
8.10 With a neat diagram explain construction and working of agitated tank.
8.11 Define impeller.
8.12 Briefly discuss on impellers.
8.13 With a neat diagram explain the working of propellers.
8.14 What is turbine? How it work?
8.15 With a neat diagram explain the working of paddle.
8.16 What is turbine? How it is used in mixing.
8.17 Define swirling.
8.18 Define vortex.
8.19 Explain the method of elimination of swirling and vortex in mixing.
8.20 With a neat diagram explain the construction and working of Airjet mixer.
8.21 With a neat diagram explain the construction and working of Jet mixer.
8.22 With a neat diagram explain the construction and working of Flow mixer.
8.23 With a neat diagram explain the construction and working of Tumbler blender.
8.24 With a neat diagram explain the construction and working of Double cone tumbling blender.
8.25 With a neat diagram explain the construction and working of Ribbon blender.
8.26 With a neat diagram explain the construction and working of Muller mixer.
8.27 With a neat diagram explain the construction and working of Kneader mixer.
8.28 With a neat diagram explain the construction and working of Banbury mixer.
8.29 With a neat diagram explain the construction and working of Barrel mixer.
8.30 With a neat diagram explain the construction and working of zig-zag mixer.
8.31 With a neat diagram explain the construction and working of Silverson mixer emulsifier.
8.32 With a neat diagram explain the construction and working of Ultrasonic emulsifier.
8.33 With a neat diagram explain the construction and working of Gas-liquid mixer.

EXERCISES

8.1 A flat blade turbine with six blade is installed centrally in a vertical tank. The tank is 1.83 m in diameter. The turbine is 0.61 m in diameter and is positioned 0.61 m from the bottom of the tank. The turbine blades are 127 mm wide. The tank is filled to a depth of 1.83 m with a solution of 50%. Caustic soda at 65.6°C which has viscosity 12 Cp and density of 1498 kg/m^3. The turbine is operated at 9 RPM and the tank is baffled.

(*i*) What power will be required to operate the mixture if $N_P = 5.8$ for Reynolds number ranging from 50,000 to 65,000?

(*ii*) If caustic soda is replaced by a rubber compound having a viscosity of 1200 Cp and density of 1200 kg/m^3. What will be the power required for this case? Assume $K_L = 65$.

❑❑❑

9

CHAPTER

Sampling, Storage and Conveying of Solids

9.1 INTRODUCTION

The action or process of taking samples of something for analysis is called sampling. In statistics, quality assurance and survey methodology, sampling is the selection of a subject of individuals from within a statistical population to estimate characteristics of the whole population. Statistics attempt to collect samples that are representative of the populations. Sampling required lower costs and faster data collection than measuring the entire population and can provide insights in cases where it is infeasible to sample an entire population.

Each observation measures one or more properties of independent objects or individuals. In survey sampling weight can be applied to the data to adjust for the sample design particularly in stratified sampling. Results from probability theory and statistical theory are employed to guide the practice.

The place where things can be kept safely when they are not being used.

Bringing of things from one place to another place is called conveying.

9.2 NEED FOR SAMPLING

The batches of grain are rarely uniform in quality even when regarded as acceptable. Pests usually occur non-randomly in stored grain. Consequently the only sure way of obtaining complete and accurate information about the grain is to carry out a total examination. This may be possible if the quantity to be examined is small, but is usually neither practical nor economical when a large quantity is involved. The choice is either not to examine the consignment at all or to take samples to obtain some information, acknowledging that anything less than a total examination is bound to affect the accuracy of the results.

The sampling can be done for many things. But, here only a few sampling processes are discussed.

(*i*) Waste-water sampling

(*ii*) Grain sampling

(*iii*) Beverage sampling

9.2.1 Waste-Water Sampling

Waste-water sampling is generally performed by one of the two methods, grab sampling or composite sampling.

Grab sampling is just what it sounds like all of the test material is collected at one time. The material being sampled is collected in a common container over the sampling period.

As such as grab sample reflects performance only at the point in time that the sample was collected and then only if the sample was properly collected. Composite sampling consists of a collection of numerous individual discrete samples taken at regular intervals over a period of time usually 24 hours. The material being sampled is collected in a common container over the sampling period. The analysis of this material collected over a period of time, will therefore represent the average performance of a waste water treatment plant during the collection period.

9.2.2 Grain Sampling

The grain sampling is important because information from the sample is used to establish the quality characteristics and the value of the grain. Therefore, it is important that proper thought and attention be given to the method of collection of sample and frequency of sample collection per unit volume of grain.

The sampling should be done at the farm as well as at the elevator to determine the quality of the grain going into storage. Knowing the grain quality permits marketing grain based on both quality and quantity, and having a representative sample permits taking the sample to several buyers for analysis and a price quote.

The sampling is also necessary for proper grain storage, management. Information from the samples, such as grain moisture content and the amount of foreign material, can be used to determine appropriate action to maintain the quality of the stored product.

9.2.3 Beverage Sampling

The beverage sampling is process used to check that a beverage is safe and that it does not contain harmful contaminants or that it is contains only permitted additives at acceptable levels or that is contains the right levels of key ingredients and that its label declarations are correct or to know the levels of nutrients present.

A beverage sample is carried out by subjecting the product for physical analysis. Analysis may be undertaken by or on behalf of a manufacturer regarding their own product or for official food law enforcement or control purposes.

9.3 STORAGE BIN

The storage bin is the smallest available unit of space in a warehouse. The storage bin therefore describe these position in the warehouse where the goods are or can be stored.

The attention must be given for the storage of solids. It is an important operation, when the material is too valuable or too soluble when exposed to outdoor ones. The following devices are used for storing solids.

(*i*) Bunker

(*ii*) Bin

(*iii*) Silo

9.3.1 Bunker

It is a large storage container for coal awaiting use or transportation. This can be either in domestic, commercial or industrial premises or on a ship or locomotive tender or coal processing plant.

Domestic coal bunker are associated with the use of coal in open fires or for solid fuel central heating. Free standing bunkers are commonly made of wood or concrete are currently sold in the market. Bunkers could also form an outhouse or be partly or fully underground. A typical bunker is shown in Fig. 9.1.

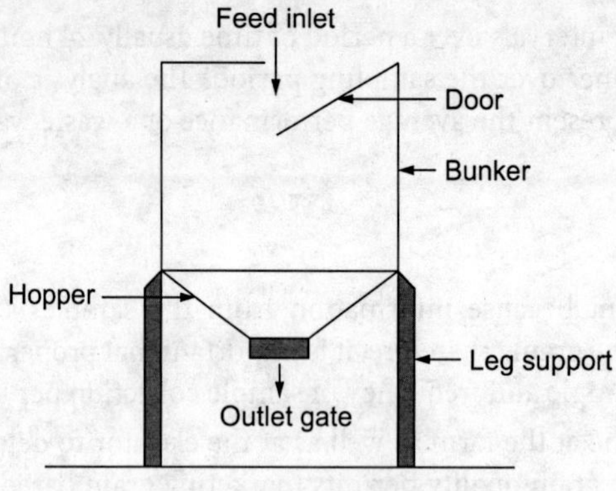

Fig. 9.1: Bunker

The material to be stored is fed from the top. As the bunker is filled with the material, the inlet door is closed and the material is discharged from the outlet gate.

9.3.2 Bin

A bin is fairly wide and not so tall. It is a cylindrical or rectangular vessel of concrete or metal. Generally it is used for storage of bulk material. A typical Bin is shown in Fig. 9.2.

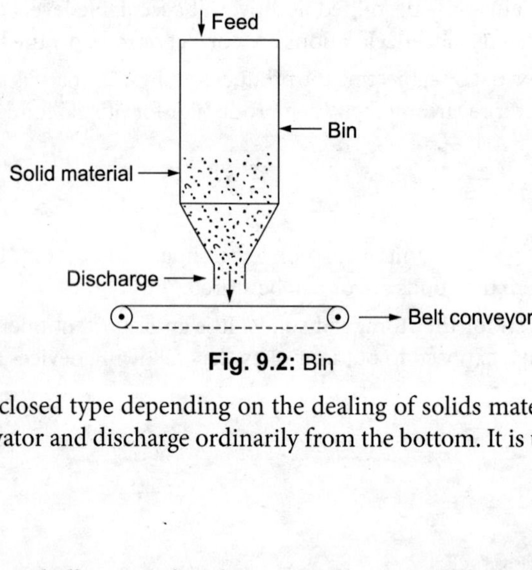

Fig. 9.2: Bin

Bins could be open or closed type depending on the dealing of solids material. This bin is loaded from the top by some kind of elevator and discharge ordinarily from the bottom. It is used in paper mill for storing Bamboo pieces.

9.3.3 Silo

A silo is a structure for storing bulk materials. Silos are used in agriculture to store grain or fermented feed known as silage. Silos are commonly used for bulk storage of grain, coal, cement, carbon black, wood chips, food products and sawdust. Three types of silos are in widespread use today. They are:

(*i*) Tower silo

(*ii*) Bunker silo

(*iii*) Bag silo

It consists of a cylindrical chamber with a maximum height 7.2 m and diameter 2.5 m. The bottom of the silo is usually conical and the top either flat or dish type. A typical silo is shown in Fig. 9.3.

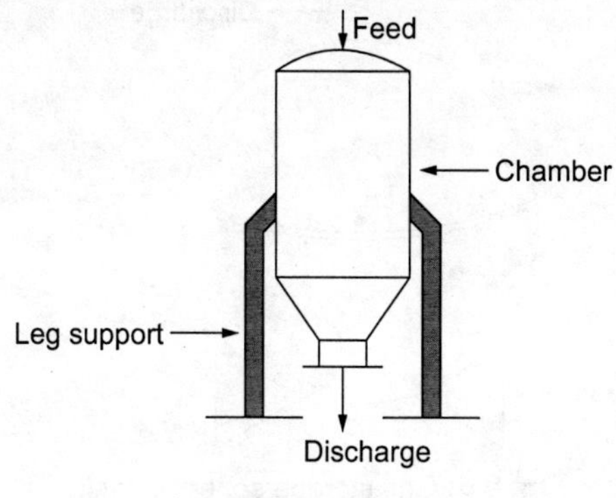

Fig. 9.3: Silo

Silo could be open or closed type depending on the dealing of solid materials. It is loaded from the top with the help of elevator and discharge from the bottom.

9.4 STORAGE OF LIQUID

Liquids are generally stored in tank. Liquid storage tanks are commonly used in industries for storing chemicals, petroleum products and for storing water in public water distribution systems.

Importance for ensuring safety of such a tanks against seismic loads cannot be overemphasized. The industrial liquid storage tanks comes in different size and shape. They can be underground, horizontal, vertical and be made from concrete, stone, fiber glass, steel or plastic. A typical liquid storage tank is shown in Fig. 9.4.

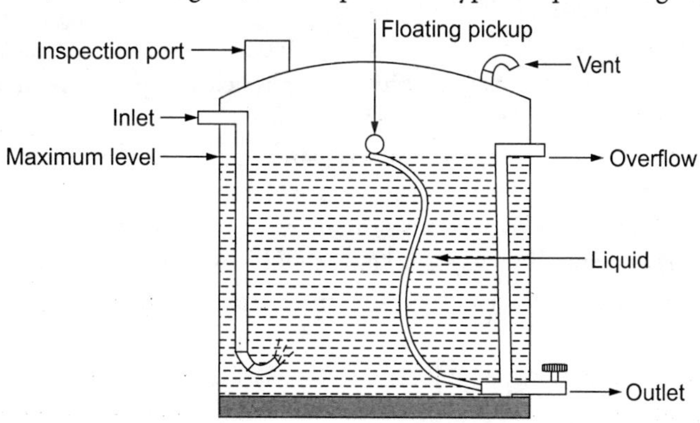

Fig. 9.4: Liquid storage tank

A floating ball is placed inside the tank along with the discharge line. As the liquid discharge from the tank, level of the liquid come down along with the floating pick up. Hence, the process continued.

9.5 STORAGE OF GAS

Spherical storage vessels are used to store below ambient temperature liquids and pressurized gases such as ammonia, propylene, LPG, butadiene etc. Most of the spherical vessel operates at low temperatures with –50°C as a lower limit. The gases are stored under pressure at a temperature lower than their liquefaction temperature.

The main advantage of the spherical tank construction is that the stress concentration in a spherical shape will be minimum while storing pressurized gases as the stress resistance will be uniform over the total surface. Another advantage is the reduced exterior surface for a given storage volume as compared to all other possible shapes. A typical gas storage spherical tank is shown in Fig. 9.5.

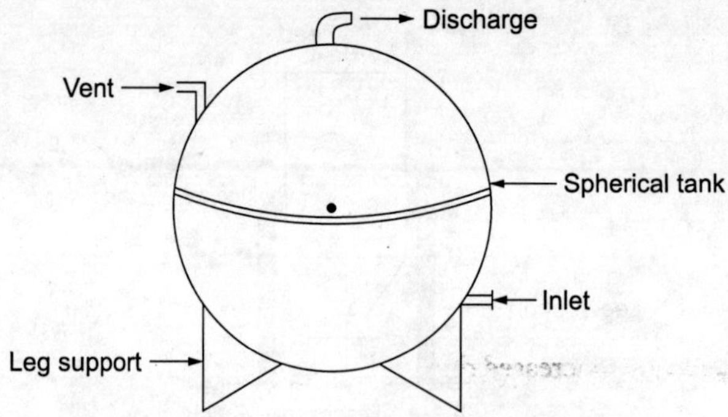

Fig. 9.5: Gas storage spherical tank

The gas is allowed to enter the tank from the bottom until it attain a required pressure while it discharge from the top.

9.6 CONVEYING OF SOLIDS

The term conveying is applied to the transportation of solids. The transportation of liquid is much simpler, cheaper and less troublesome than handling solids. In many operation solids are handled in a finely divided state, so that they remain suspended in a stream of fluid. However, it may be necessary to transport solids as such. This until operation is important in the storage and handling of raw materials, finished products and also for packed goods. If raw materials are not moved as required by production schedule, the manufacturing process will be slowed down. In addition, machine time and operation time will be wasted. As a result in processes inventories will be increased.

Improper handling and storage can lead to damage, outdated and loss of materials. Improper handling of granules, strips, bottles and chemical can place employees in physical danger. This can often lead to employees frustration and reduced morale due to constant production delays.

In today's world of automation, solid conveying has gained a greater importance. It is often necessary to prevent the contact of material with the persons, in order to avoid contamination. Similarly, working personnel should not be exposed to poisonous and obnoxious substances. Such a closed circuit manufacturing operation calls for automatic handling systems. Similarly, discharge should also be made automatic.

The conveyors are classified into five major groups. These are:

(*i*) Belt conveyor
(*ii*) Apron conveyor
(*iii*) Screw conveyor
(*iv*) Pneumatic conveyor
(*v*) Chain conveyor
(*vi*) Flight conveyor
(*vii*) Bucket conveyor

9.6.1 Belt Conveyor

A belt conveyor is a continuously moving strip of rubber or metal which is used in factories for moving objects along so that they can be dealt with as quickly as possible.

The basic element of the belt conveyor are:

(*i*) Belt and tightening system
(*ii*) Belt drive and power supply
(*iii*) Roller supports for belt

(*iv*) Feeding arrangement

(*v*) Discharge arrangement

The belt conveyor consists of a core or carcass of several piles of cotton duck, each layer is impregnated and bonded with rubber. The carcass is then covered with a thin layer of rubber that binds the piles together. Both the ends of the belt are joined together, so that the belt can revolve continuously. The belt may run horizontally or slightly inclined from the feed point.

If the belt is too thin for its width, it will sag between the idlers. If the belt is too thick, it does not trough properly.

Sometimes, the length of belt is increased due to a variety of factors such as temperature and humidity in the atmosphere. In such cases, it is necessary to keep the belt taut. Hence, tightener are installed to maintain in even-tension on the belt under all conditions. The simplest take-up consists of a cast iron bed with a travelling block moving along a screw. A typical Belt conveyor is shown in Fig. 9.6.

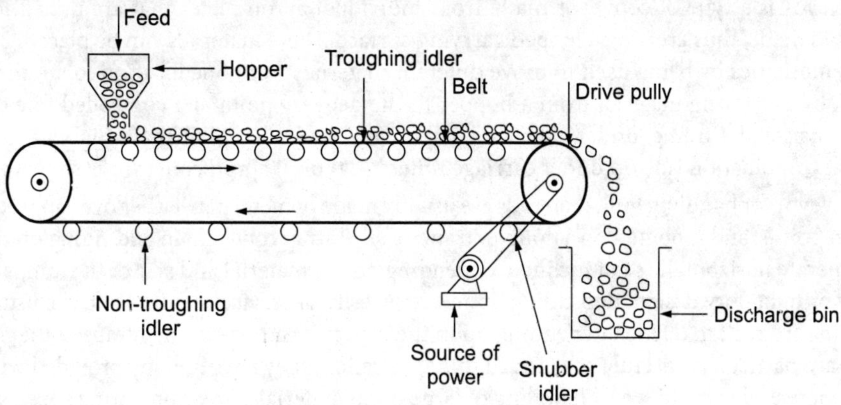

Fig. 9.6: Belt conveyor

The belt is allowed to move with the help of a drive. The simplest belt drive is a bare steel pully actuated with some source of power. The drive of the belt conveyor is fixed at the discharge end. There should be appropriate contact between the belt and pully. Introducing snubber idlers just below the pully can increase the contact.

The belt is supported by the roller, which are arranged on a shaft. These are called idlers. The idlers are carried on bush bearings lubricated with grease-cups. The idler are generally troughed to depress the belt at the centre and rise at the edges. The different belt support are shown in Fig. 9.7.

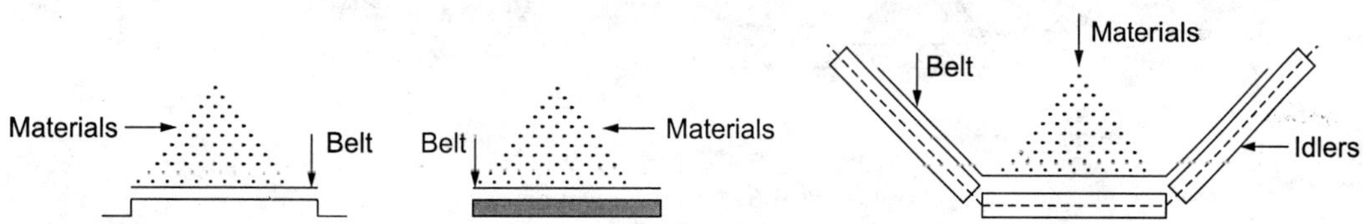

(*a*) Flat belt on continuous plate (*b*) Flat belt on idler (*c*) Troughed belt on idler

Fig. 9.7: Belt support

This permits the belt of a given width to carry more material per linear meter without spillage. The belt returns on ordinary non troughing rolls.

A hopper is placed at one end in such a manner to load the feed at the centre of the belt. Cleaning devices such as revolving brush and rubber scrapper blades are placed in the path of the belt in order to clean, if the material stick to the belt. Provision is made at the other end for discharging.

When the power is applied to a conveyor motor, the belt drive is set in rotation. As the pulley moves, the belt begins to travel in forward direction, because of its close contact with the pulley. Snubber indlers also help in maintaining the close contact between the pulley and belt.

The material to be transported is loaded on its centre of the belt with the help of a hopper. The material travel along with the belt. The belt moves forward on troughing idlers. This permits the belt to carry more material per linear meter of the belt without spillage. The amount of feed delivered depends on the width and speed of the belt. At the receiving end the material may be unloaded manually or mechanically.

Once the material is unloaded, the returning belt is carried on lighter non-troughing rolls. The belt return to the point of feeding. The belt conveyor is used in coal mine to transfer the coal at thermal power station for generation of power. It is very expensive.

9.6.2 Apron Conveyor

An apron conveyor is a type of conveyor made from individual apron plates that are linked together with hinges on its underside, thus creating a looped carrying surface where materials can be placed. It plays a vital part in several industries by being used to move different materials from one location to another. The apron conveyor work by conveying material from a hopper to the delivery point at a controlled rate of speed. The apron feeder consists of a frame, under carriage components, head shaft, tail shaft, flights and a drive. The chain and flight combination ride on under carriage rollers. A typical apron conveyor is shown in Fig. 9.8.

Apron conveyor for handling bulk materials are usually made of metal pan, either overlapping or hinged at the articulation points, and mounted on a double strand of steel strap roller chain and riding on single flanged rollers. They operate horizontally or at inclined, depending on the material and size of the lumps. They can be made in heavy or light duty design, depending upon the severity of service required and are usually made up in widths varying from 12 to 60 inch depending upon the capacity and size of the lumps of the material to be handled. They are particularly useful as feeder. In this application, they are generally provided with stationary side plates for increased capacities and handling of large lump materials. Their operating speed as feeders are slow, but as conveyors, they can run safely at upto 100 rpm.

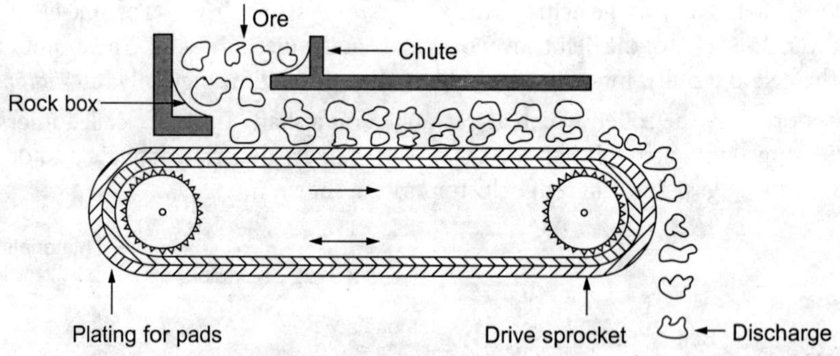

Fig. 9.8: Apron conveyor

9.6.3 Screw Conveyor

A screw conveyor is a mechanism that uses a rotating helical screw blade called flighting, usually within a tube to move granular materials. They are used in many bulk handling industries. Screw conveyors in modern industry are often used horizontally or at a slight inclined as an efficient way to move semi-solid materials, including food waste, wood chips, aggregates, cereal grains, animal feed, boiler ash, bone meal, municipal solid waste and many others.

The first type of screw conveyor was the Archimedes screw used since ancient time to pump irrigation water. The basic element of the screw conveyor are:

(*i*) Trough system

(*ii*) Flights and power supply

(*iii*) Feeding arrangements

(*iv*) Discharge arrangements

It consists of a U-shaped vessel in which the material is enclosed during conveying. The trough is usually made of a sheet of steel and available in standard lengths. If a trough of more length is required, then these can be joined together. A typical screw conveyor is shown in Fig. 9.9.

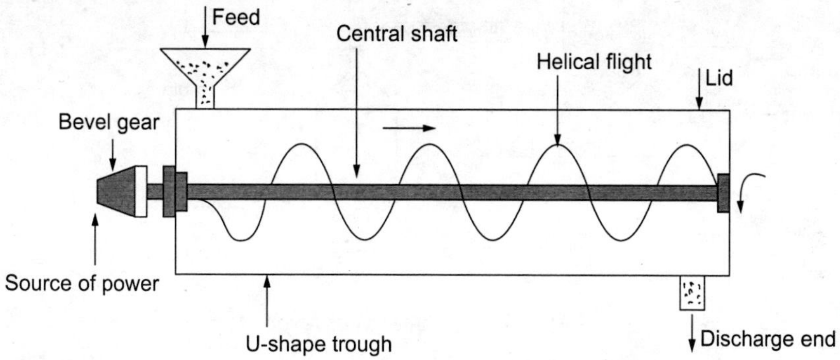

Fig. 9.9: Screw conveyor

Screw conveyor usually consists of a spiral blade, which is called flight. The simplest form of the flight is the sectional conveyor. It is made up of a single long ribbon that is twisted and wrapped into a spiral shape. These standard flights are welded to the central shaft. The conveyor element is also suspended with the help of hangers along the length of the trough. Usually one hanger is used for each flight section. Hangers contain bushing of white cast iron so that lubricant need not to be used.

The drive end and discharge end are known as box-ends. At the drive end, shaft is connected to the drive by bevel gears. Power is transmitted through the shaft to the flight. The shaft is rotated around an axis and remained nearer to the bottom of the trough. The feed is normally introduced by plain spouts, which gives uniform flow of material. At the discharge end, arrangements are made to receive the material using open and trough.

As the power is supplied, the drive rotates the shaft through the bevel gears. The screw element is set in motion. The flight is rotated round, an axis and remained nearer to the bottom of the trough. Feed is normally introduced by plain spouts. The material will be trapped between the gaps of the spiral blades. As the flight moves, the material also moves forward along the path of the sections. The material is received using open-end trough or open bottom trough.

The screw conveyor is used for transporting finely divided solids or pasty solids. The main dis-advantage is that it operates at relatively low rotational speeds.

9.6.4 Pneumatic Conveyor

It is the method of transporting materials using air as conveying medium. When a high velocity air is passed through a bed of solid particles, the individual particles are dragged with the air. Finally they are suspended in the air. Such a system is known as fluidized bed, which is described as a condition of fully suspended particles. In this condition, the suspension behave like a dense fluid and can be transported from one point to another in processing plants. At the discharge end, the air is separated and the solids are recovered.

The blower is connected to the conveying system. An air slide consists of a chute with a porous base through which air passes. The path of material travel may vary from a few meter to several hundred meters. Conveyor mechanism can be operated for horizontal and vertical distances. The material supply is connected to the air slide through a rotary feeder valve. The other end of the pipeline is connected to a cyclone separator. A typical Pneumatic conveyor is shown in Fig. 9.10.

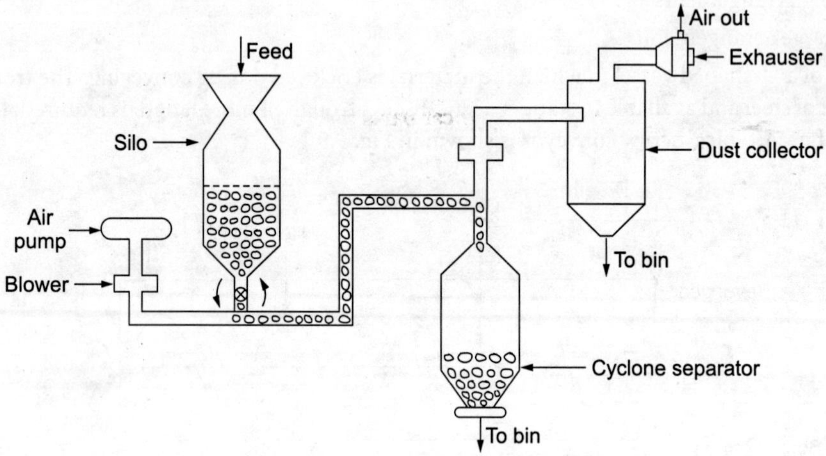

Fig. 9.10: Pneumatic conveyor

The blower produce air at a pressure about 7.0 kPa. This air is passed through the air slide at the rate of 1.5 meter per minute. During this process, the rotary feeder valve rotates and allows the feed to enter the pipe line. These solids are suspended into a stream of air in a fluidized state until it reaches the receiving end. The material is then admitted to cyclone separator to remove large particles and fine particles, if any. The large particles are collected into the bin. The fine particles are removed in a bag filter. The air can be recycled to the blower inlet. This system is used for free flowing materials.

For the design and satisfactory operation of pneumatic conveying, it is necessary to know about properties of material such as density, particle size distribution, cohesiveness, angle of friction, angle of repose and fluidization characteristics.

9.6.5 Chain Conveyor

Chain conveyor consists of chain and chain attachment. The chain element is made of malleable cast iron or stainless steel. These links are so cast that they can be assembled and detached without use of tools. Chains are endless and move continuously. The material is loaded into suitable container and placed on the chains. These are usually constructed on the job. A typical chain conveyor is shown in Fig. 9.11.

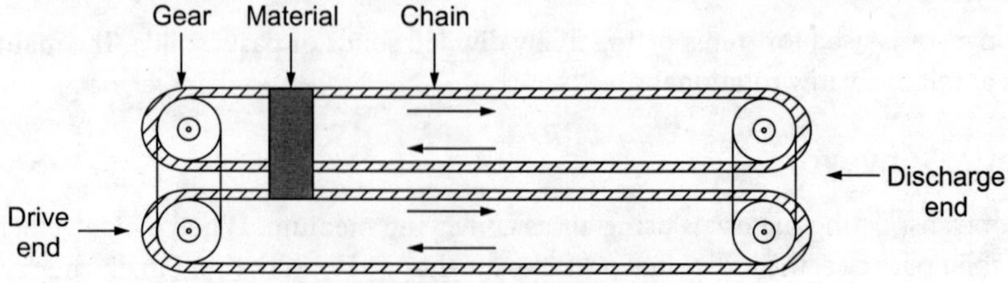

Fig. 9.11: Chain conveyor

Chain conveyors utilize a powered continuous chain arrangement, carrying a series of single pendants. The chain arrangement is driven by a motor, and the material suspended on the pendants are conveyed. The chain conveyors are used for moving products down an assembly line around a manufacturing facility.

Chain conveyors are primarily used to transport heavy unit loads. For example pallets, grid boxes and industrial containers. These conveyors can be single or double chain strand in configuration. The load is positioned on the chains, the friction pulls the load forward. Chain conveyors are generally easy to install and have very minimum maintenance for users.

Many industry sectors use chain conveyor technology in their production lines. The automotive industry commonly uses chain conveyor systems to convey car parts through paint plants. Chain conveyors also have widespread use in the white and brown good metal finishing and distribution industries. Chain conveyors are also used in the painting and coating industry, this allows for easier paint application. The products are attached to an above head chain conveyor, keeping products off of the floor allows for higher productivity levels.

9.6.6 Flight Conveyor

The flight conveyor consists of one or two endless chains passing through a trough or set of guides. The chain have plates of wood or steel called flights attached at regular intervals. The flights are shaped to fit the troughs. The chains pull the flights and the material along the trough and passes over sprockets at the end of the run. One of the spockets acts as the drive. In an elaborate form flight conveyor is supported by two chains of roller pintle. A typical flight conveyor is shown in Fig. 9.12.

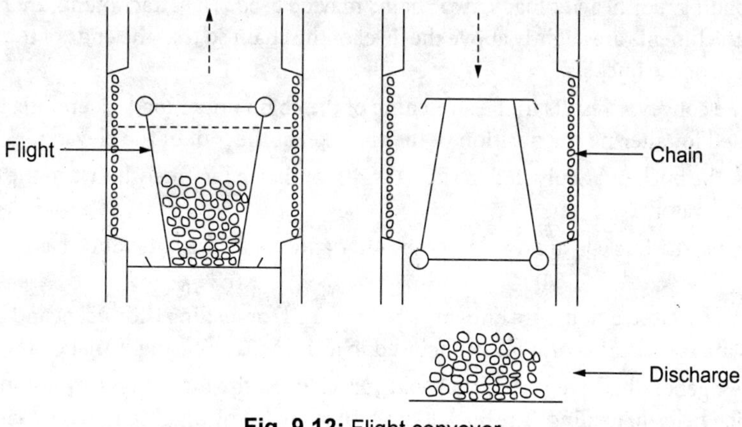

Fig. 9.12: Flight conveyor

Flight conveyors with speeds of 30 meter per minute are common, but the speed may ranges from 8 to 60 meter per minutes.

It is widely used for loose material that is non-abrasive, such as grain, food waste, garbage, sawdust, chips and coal. It is not suitable for friable abrasive material such as clinkers, gravel and crushed ore.

It is simplest and cheapest type of conveyor. The main dis-advantage is that it needs heavy power.

9.6.7 Bucket Conveyor

The buckets conveyor's are used for loading the materials. The buckets are made of cast iron or stamped steel and available in many forms. Flatter buckets are used for materials that are sticky. Heavily stamped steel buckets are employed for large lumps. For example coal and crushed stone. The buckets are attached to belts or chains. The chains are endless and buckets are fixed between two chains. A typical bucket conveyor is shown in Fig. 9.13.

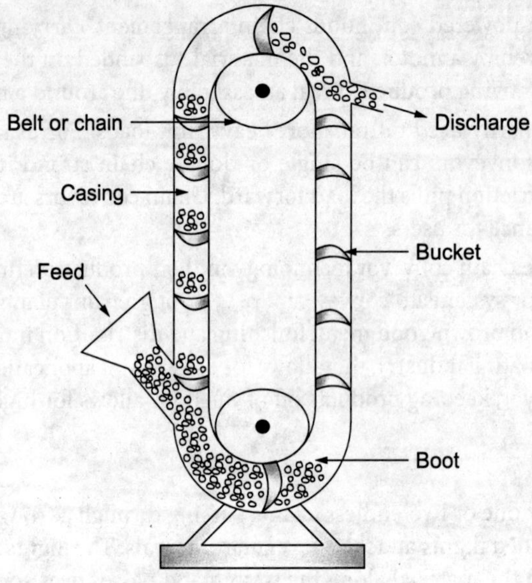

Fig. 9.13: Bucket conveyor

The chains are generally long pitch, straight endless type. The buckets are spaced to prevent interference in loading and unloading. For heavier loads, two chains may be used. The attachments are riveted to the ends of the chains. The attachments are slightly above the level of the chain so that the ends of the sprocketed booth do not strike the back of the buckets.

The weight of the conveyor results in the stretching of the chain under load. Therefore, tightening devices are generally provided for altering the position of the sprocket at the foot of the elevator.

In vertical runs, the buckets are pivoted in such a way that they hang freely between the chains. Therefore, conveyor acts as an elevator.

In horizontal runs, the buckets are overlap each other and the feed is introduced as a continuous stream of material.

As the foot of the elevator, a structure known as boot is used for feeding the bucket and also for tightening the chains. Buckets are sometimes completely enclosed in a casing. The casing is made of wood or steel sheet.

The elevators are generally driven from the head sprocket. At the foot of an elevator, there is a structure known as boot, which helps in feeding. The buckets are normally fed by digging into the materials particularly in case of loose material and granular solids. The buckets maintain their carrying capacity by gravity, but they can convey the material horizontally or vertically or any desired path. Loading can be one by the flow of material through a chute into the buckets. Buckets may be readily discharged at any point by a tripping device, causing each bucket to turn through 90°. The type of discharges are continuous, positive and centrifugal.

The bucket conveyors are used in transporting coal, crushed stone and grains. It is very flexible. The main dis-advantage is that it needs heavy power.

REVIEW QUESTIONS

9.1 Define sampling.
9.2 Explain the need sampling.
9.3 Explain the method of waste-water sampling.
9.4 Explain the method of grain sampling.
9.5 Explain the method of beverage sampling.
9.6 What do you mean by storage?
9.7 Define bunker.
9.8 Explain the method of storing solids in bunker.
9.9 Define bin.
9.10 Explain the method of storing solids in bin.
9.11 Define silo.
9.12 Explain the method of storing solids in silo.
9.13 With a neat diagram explain the process of loading and unloading of material from the bunker.
9.14 With a neat diagram explain the process of loading and unloading of material from the bunker.
9.15 With a neat diagram explain the process of loading and unloading of material from the silo.
9.16 With neat diagram explain the process of loading and unloading of liquid from the liquid storage tank.
9.17 With a neat diagram explain the process of loading and unloading of gas from the gas storage spherical tank.
9.18 Define conveying.
9.19 Classify the conveyors.
9.20 With a neat diagram explain the construction and working of Belt conveyor.
9.21 With a neat diagram explain the construction and working of Apron conveyor.
9.22 With a neat diagram explain the construction and working of screw conveyor.
9.23 With a neat diagram explain the construction and working of pneumatic conveyor.
9.24 With neat diagram explain the construction and working of chain conveyor.
9.25 With a neat diagram explain the construction and working of flight conveyor.
9.26 With a neat diagram explain the construction and working of Bucket conveyor.

❑❑❑

10

CHAPTER

Magnetic Separations

10.1 INTRODUCTION

It is the process of separating components of mixtures by using magnet to attract magnetic materials. The process that is used for magnetic separation detaches non-magnetic material with those that are magnetic. This technique is useful for not all, but few minerals such as ferromagnetic and paramagnetic materials. Not all metals are magnetic. For example gold, silver and aluminium.

A large diversity of mechanical means are used to separate magnetic materials. During magnetic separation, magnets are situated inside two separate drums which bear liquids. Due to the magnets, magnetic particles are being drifted by the movements of the drums. This can create a magnetic concentrate.

Michael Faraday discovered that when a substance is put in a magnetic environment, the intensity of the environment is modified by it. With this information, Michael Faraday discovered that different materials can be separated with their magnetic properties.

Magnetic separation can also be used in electromagnetic cranes that separate magnetic material from scraps and unwanted substance. This explains its use for shipment equipment and waste management. Unwanted metals can be removed from goods with this technique. It keeps all materials pure. Recycling centres use magnetic separation often to separate components from recycling, isolate metals and purify ores. Overhead magnets, magnetic pulleys and magnetic drums were the methods used in the recycling industry.

Magnetic separation is also useful in mining iron as it is attracted to a magnet.

Another application not widely known but very important, is to use in process industries to remove metal contaminants from product streams. This takes a lot of importance in food and pharmaceutical industries.

Magnetic separation is also used in situations where pollution needs to be controlled, in chemical processing as well as during the beneficiation of non-ferrous low grade ores.

Magnetic separation is also used in the following industries such as dairy, grain, milling, plastics, food, chemical, oil, textile and many more.

There are many methods available for magnetic separations. But, here only the following methods are discussed:

(*i*) Magnetic head pulley separator

(*ii*) Magnetic drum separator.

(*iii*) Ball Norton magnetic separator.

(*iv*) Electrostatic separator.

(*v*) Heavy media separator.

10.1.1 Magnetic Head Pulley Separator

The Magnetic Head Pulley Separator is an industrial strength magnet which provides a dependable means of tramp metal extraction on conveyor belt applications. The permanent forces protect the main equipment, increase uptime, reduce maintenance costs and ensure highest product purity.

A magnetic head pully separator is a self cleaning magnet which removes foreign tramp metals from bulk materials on a belt conveyor. The magnetic field adheres ferrous contaminants to the belt as clean product passes over the head pulley. Tramp metals from the underflow can be safely captured in a bin to be discarded. A typical Magnetic Head Pulley Separator is shown in Fig. 10.1.

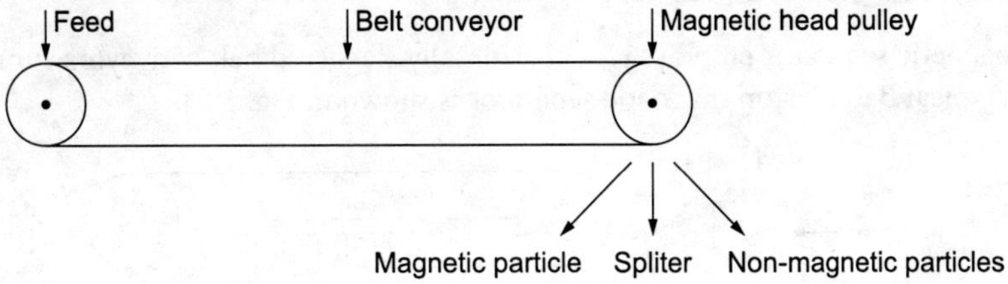

Fig. 10.1: Magnetic head pulley separator

A magnetic take off rail conveyors captured tramp metal further down the return side of the conveyor. The take off rail paired with the Magnetic Head Pulley allows for a more convenient and smoother release of retained tramp metal reducing belt wear while discharges in a suitable location.

Magnetic pulleys are used for reliable ferrous metal recovery in a wide variety of industries such as recycling, aggregate and waste recovery operations. It is critical to select the correct magnetic pulley for the operation. Standard applications typically require deep reach ceramic magnet while high intensity neo rare earth magnets are used in more aggressive applications.

10.1.2 Magnetic Drum Separator

The magnetic drum separator is a device used to separate the components of mixtures by using a magnet to attract magnetic materials. The process that is used for magnetic separation detaches the non-magnetic material with those that are magnetic. This can create a magnetic concentrate.

Magnetic drum separator consists of a rotating drum incorporating stationary magnet assembly. The magnets are covers approximately 165° towards discharge side of the drum. A typical magnetic drum separator is shown in Fig. 10.2.

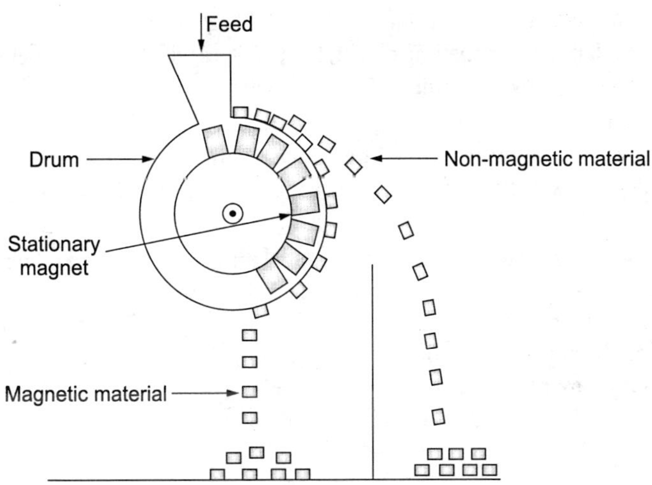

Fig. 10.2: Magnetic drum separator

The feed is admitted at the top and is allowed to fall on the rotating drum. The non-magnetic material is discharged in normal manner while the magnetic materials adhere to the drum and falls of underside when the drum loses the contact of magnetic assembly.

The common application of magnetic drum separator are it keeps all materials pure. Recyclinbg centres use magnetic separation often to separate components from recycling, isolate metals and purify ores. Magnetic separation is also used in the following industries such as dairy, grain and milling, plastics, food, chemical, oils, textiles and many more.

10.1.3 Ball Norton Magnetic Separator

The Ball Norton magnetic separator employs two horizontally staggered belt conveyors running parallel one above the other. A typical Ball Norton magnetic separator is shown in Fig. 10.3.

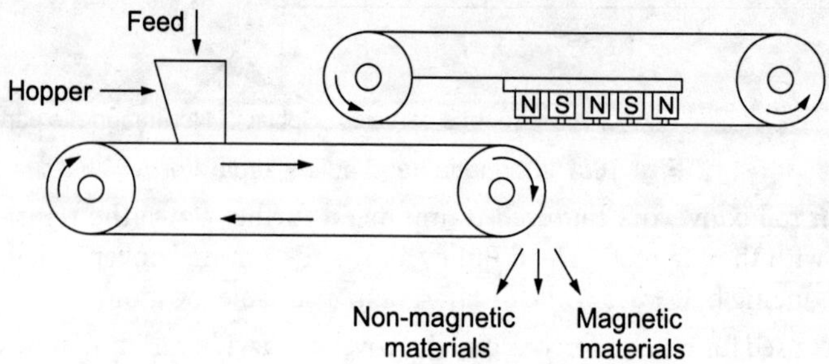

Fig. 10.3: Ball Norton magnetic separator

The hopper is provided for feeding the material to the lower belt and stationary magnet assembly is incorporated in the upper belt conveyor near discharge end. The materials to be separated is fed to the lower belt in the form of thin sheet and is conveyed under the second belt where it is subjected to magnetic field. The non-magnetic material is discharged in a normal manner while the magnetic material adheres to the lower side of the upper belt. The magnetic materials is thus, carried some distance away from discharge of non-magnetic materials and is then discharged from the belt in separate compartment when the belt loses the contact of magnet assembly. Hence, the magnetic material is separated from the non-magnetic materials.

10.1.4 Electrostatic Separator

The electrostatic separation is a process that uses electrostatic charges to separate crushed particles of material. When a mixture of particles falls past a repelling object, the particles with the correct charge fall away from the other particles when they repelled by the similarly charged object.

An industrial process used to separate large amounts of material particles, electrostatic separating is most often used in the process of sorting mineral ore. This process can help remove valuable material to purify a substance. In mining, the process of crushing mining ore into particles for the purpose of separating minerals is called benefication.

Generally, electrostatic charges are used to attract or repel differently charged material. When electrostatic separation uses the force of attraction to sort particles, conducting particles stick to an oppositely charged object, such as a metal drum, thereby separating them from the particle mixture. When this type of benefication uses repelling force, it is normally employed to change the trajectory of falling objects to sort them into different places. This way, when a mixture of particles falls past a repelling object, the particles with the correct charge fall away from the other particles when they are repelled by the similarly charged object.

An electric charge can be positive or negative. Objects with a positive charge repel other positively charged objects, thereby causing them to push away from each other, while a positively charged object would attract to a negatively charged object, thereby causing the two to draw together.

An electrostatic separators differences in the electrical properties of the materials are explored to effect the separation.

It consists of a grounded rotating drum, a hopper for feeding the solids, an active electrode situated at small distance from the drum and collecting bins. A typical electrostatic separator is shown in Fig. 10.4.

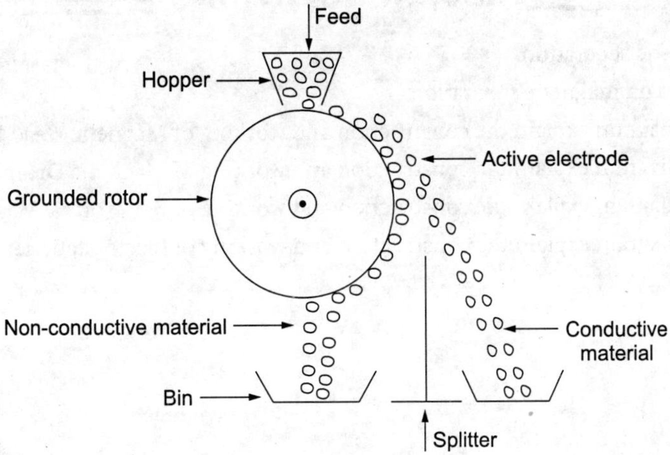

Fig. 10.4: Electrostatic separator

The solids are fed on to a rotating drum, either charged or grounded from the hopper. The conductive particles within very short periods of time will assume the potential of the rotating drum, which is opposite to that of the active electrode and hence, get attracted towards the active electrode. The non-conductive particles fall down straight under the influence of gravity and are collected in a separate bin.

10.1.5 Heavy Media Separation

Heavy media separation is also known as dense media separation or sink float processing is an adaption of the common laboratory procedure for separating solids of differing specific gravity by immersing them in a heavy liquid of specific gravity intermediary between those of the solids, thereby causing the lighter particles to float while the heavier sink. However, in heavy media separation, the parting solid is produced by dispersing relatively fine grained solids of a high specific gravity in ... and maintaining this pulp in suspension by light agitation. The method is very active and can be used to separate solids with difference in specific gravity as little as 0.005. It often the only process needed for the removal of deleterious wastes char-coal. The method is used extensively for beneficial ore mineral and it is finding increasing use for the processing of shredded automobile scrap and for the recovery of particles from solids municipal wastes.

These media separation may be used to produce either a finished concentrate or a feed for subsequent processing. In the later case, it provides a low cost means to produce a significant amount of essentially barren waste at a coarse size.

The float plants are usually custom designed for each individual application. However, for the benefication modular units are available. For most large mineral processing beneficiations the plant will be permanently located for easy access of feed and disposal of unwanted ore material but for smaller coal and aggregate operation the plants are often constructed with the temporary base for relocation when the deposits have been depleted.

The response of any given feed to sink float processing can be accurately established in the laboratory by testing with various heavy liquids. The liquid generally used for this purpose ethylene iodide. These homogenate hydrocarbons are mutually miscible, which enables the preparation of almost any pulp density attainable in a commercial plant. Heavy liquid test work provides the basis for specifying the optimum screen size for the preparation of the feed.

REVIEW QUESTIONS

10.1 Define magnetic separation.

10.2 Briefly discuss on magnetic separation.

10.3 With a neat diagram explain the construction and working of Magnetic Head Pulley separator.

10.4 With a neat diagram explain the construction and working of Magnetic Drum separator.

10.5 With a neat diagram explain the construction and working of Ball Norton Magnetic separator.

10.6 With a neat diagram explain the construction and working of Electrostatic separator.

❑❑❑

11

CHAPTER

Jigging

11.1 INTRODUCTION

A jig is a mechanical device used for the separation of materials of different specific gravities by pulsating the stream of liquid usually water flowing through a bed of materials resting on a screen.

Jigging is a method of separation of materials of different specific gravities by pulsating the stream of liquid usually water flowing through a bed of materials resting on a screen.

Jigging is a process of gravity concentration where solids are separated based upon difference in behavior of particles through moving fluid which in turn depends upon specific gravities. Separation of solids of different specific gravities is achieved by pulsation of liquid flowing through a bed of solids on screen. The liquid pulsates or jigs up and down causing up and causing heavy material to work down at the bottom of the bed and lighter material to rise at the top.

Jigging is one of the oldest processes used for concentrating heavy material from the light materials. It is commonly employed for coarse material that is in the size range 20 mesh and coarser and where there is a remarkable difference between the effective specific gravity of the valuable and the waste material.

These devices are simple in operation, consume very large quantities of water and have high tailings losses on metallic ores. The jigs are used mostly to treat iron ores, few lead-zinc ores. It is widely employed for the concentration of coal. In this chapter the following methods are discussed.

(*i*) Hydraulic jig

(*ii*) Froth floatation

11.1.1 Hydraulic Jig

Hydraulic jig operates by providing a very short time for material to settle so that the particles do not attain their terminal falling velocities. Therefore, it is suitable for separation of wide size range into their constituents.

It consists of a tank of rectangular cross-section with a tapered bottom. The tank is divided into two portions by a vertical baffle. In one compartment the plunger is incorporated. It operates in a vertical direction giving pulsating motion to the liquid. In the other compartment the screen is incorporated. The separation is carried out over this screen. It is provided with bottom discharge connection for removal of small particles of heavy material and gates at the side of jig for removal of particles settled on the screen and for overflow. A typical Hydraulic jig is shown in Fig. 11.1.

The material to be separated is fed over screen and is subjected to a pulsating action by oscillating with the help of reciprocating plunger. During the downward stroke the water inlet is closed and particles on the screen are brought into suspension and they segregate according to their size and density such that the dense material is collected near the bottom of the screen.

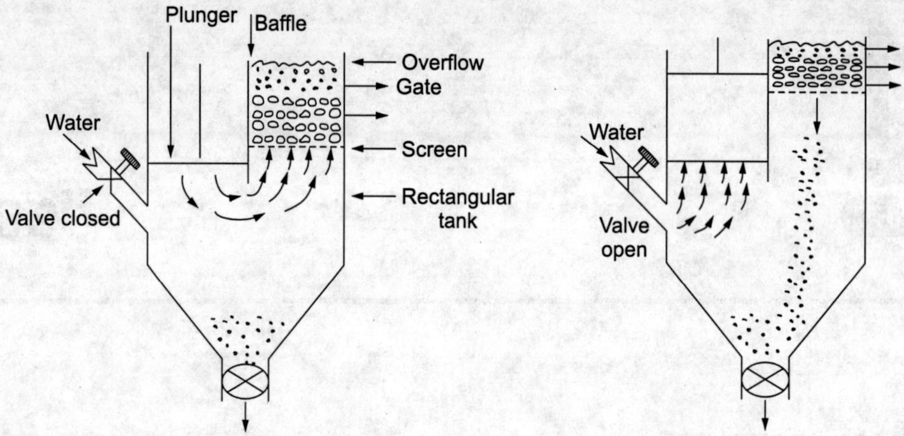

(*a*) Downward stroke (*b*) Upward stroke

Fig. 11.1: Hydraulic jig

During the upward stroke of the plunger, the input water is taken into the jig and there is no net flow through the bed of solids. The very small particles of dense material will pass through the screen and are collected at the bottom of the jig. Small particles of less dense material are remove through the overflow. Material retained on screen is removed through the gates. Generally, four fractions are obtained.

11.1.2 Froth Floatation

Froth floatation is an operation in which one solid is separated from other by floating one of them on the surface of the fluid. In Froth floatation, the separation of feed mixture depends upon the differences in the surface properties of the materials. This technique is commonly used in mineral dressing. Mineral dressing refers to the method of treating ores at or near the mine site to produce one or more concentrates of the valuable minerals and a tailing composed of waste or less valuable minerals.

Froth flotation is used for treating the metallic ores that are finer than 48 to 65 mesh or coal and certain non-metallics that are finer than 10 to 48 mesh. It is not possible to treat the coarser feed by froth floatation as same cannot be suitably mixed and suspended by the floatation cell.

Separation of solids using floatation methods depends on differences in the surface properties of the hydrophobic or hydrophilic materials. If a mixture is suspended in an aerated liquid, air bubbles will tend to adhere preferentially to one of the constituents. The one which is more difficult to wet by liquid and its effective density will be reduced to such an extent that will rise to the surface and float over the liquid, while the material which has affinity for liquid get surrounded by liquid and it will simply sink, thus achieving the separations.

To improve the floatation ability the following material may be added to the suspended solids:

(*i*) Promotors/Collectors
(*ii*) Modifiers
(*iii*) Frothing agents

11.1.2.1 Promotors/Collectors

The function of promoters in flotation is to increase the floatability of minerals in order to effect their separation from the undesirable mineral fraction commonly known as gangue.

Almost all the minerals and inorganic solids are hydrophilic, as the surface of these solids are easily wetted by water. Hydrophilic solids are unfloatable as air bubble do not surround or cover them to form particle bubble aggregate. However, these solids can be made hydrophobic with the help of reagents known as promoters or collectors. The promoters or collectors are materials which selectively render the desired particles air-avid and water repellent by getting absorbed on their surfaces.

The typical promoters are sodium ethyl xanthate (NaS · S · C · C_2H_5) and dithiophosphate (RO · RO · P · S · SNa). Collectors are materials which form surface film on the particles. The commonly used collector is pine oil.

11.1.2.2 Modifiers

In many cases, it is necessary to modify the surface of the other constituent for preventing the adsorption of promoters or collectors on the surface. It is an essential requirement of floatation technique to produce a froth of sufficient stability which can retain the particles of one constituent on the surface so that they can be discharged from the overflow. The froth should not be so stable that it is difficult to break in the later stage. The most effective modifiers are:

(*i*) pH modifiers include lime (CaO), caustic soda (NaOH), sulfuric acid (H_2SO_4) and hydrochloric acid (HCl).

(*ii*) Anionic modifiers include phosphates, silicates and carbonates.

(*iii*) Organic modifiers include dextrin, starch and glue.

11.1.2.3 Frothing Agents

Apart from the collectors/promoters and modifiers frothing agents are also used for the production of a froth of sufficient stability to retain the particles of one constituent on the surface. It is an essential requirement of floatation technique to produce a froth of sufficient stability which can retain the particles of one constituent on the surface so that they can be discharged from the overflow. The froth should not be so stable that it is difficult to break in the later stage. The most effective frothing agents are liquid soaps, pine oils, cresols and alcohols in the range between amyl and octyl alcohols.

The valuable concentrates from froth floatation may be either the froth product removed from the top or from the underflow product. In case of metallic sulfide ores of copper, zinc, nickel and lead, native gold and silver, the valuable product collects in froth and removed from the top. In glass-sand floatation, iron bearing minerals collect in froth while valuable product is removed as underflow product.

The mechanically agitated cell consists of a tank having a circular cross-section. It is provided with an agitator which violently agitates the pulp. The air from the blower is introduced into the system through the downpipe surrounding the impeller shaft. The bottom of the tank is conical and is provided with discharge from tailings. An overflow is provided at the top for mineralized froth removal. A typical lab model froth flotation cell is shown in Fig. 11.2.

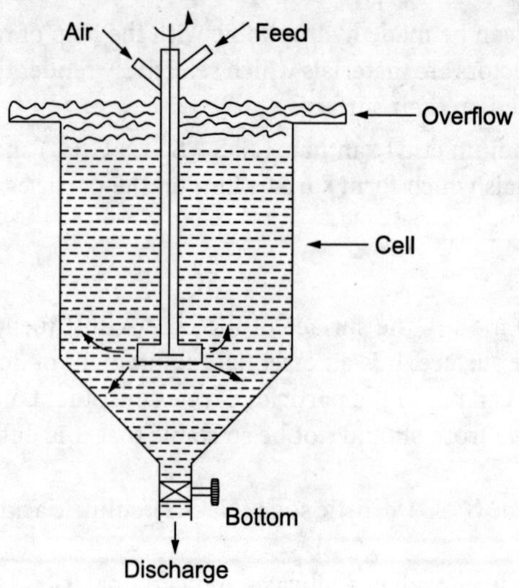

Fig. 11.2: Lab model froth floatation cell

Water is added into the cell and material is fed. The promoter and frothers are added. The agitations are given and air is passed in the form of fine bubbles. Air avid particles due to reduction in their effective density, will rise to the surface and be held in the froth before they are discharged from the overflow. Hydrophilic particles will sink to the bottom and removed from the discharged as tailings.

D R Denver Flotation Cell: The cell that is most widely used today in sulfide, coal and non-metallic flotation operations in the western hemisphere are the Fagergren, the D R Denver and the agitator flotation cell. These cells provide mechanical agitation and aeration by means of a rotating impeller on a upright shaft. A typical D R Denver flotation cell is shown in Fig. 11.3.

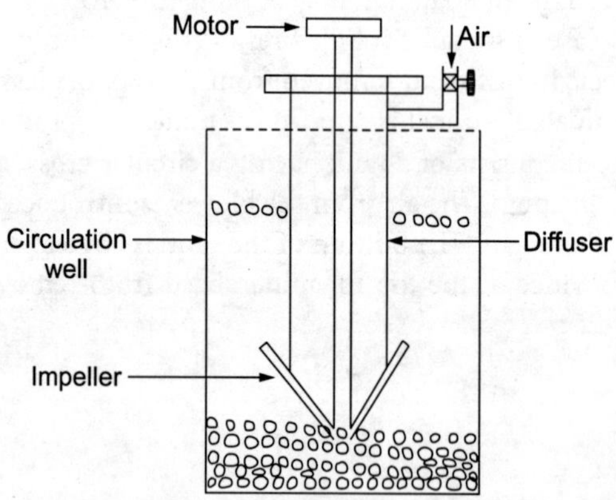

Fig. 11.3: D.R. Denver flotation cell

In D R Denver flotation cell the pulp enters from the top of the recirculation well, while low pressure air enters through the air pressure. The pulp and air intimately mixed and thrown outwards by the stationary diffuser. The collector coated mineral particles adheres to the rising bubbles and are carried to the top of the cell to be removed in the froth products.

Flotation Circuit: The ores must be ground to a point complete or nearly complete aeration. Even though this might possibly be accomplished by coarse crushing, grinding finer than 10 mesh necessary in all cases and to finer than 48 mesh in most cases prior to flotation. Grinding is done in closed circuit with classifiers. In many instances, superior flotation are obtained by conditioning the ore with the reagents before the flotation step oily ore collectors are some time added to the grinding circuit to ensure dispersion.

The proper selectivity, a definite contact time is some time required between reagent and this usually secured by mixing the reagent and ore pulp in a conditioner consisting a cylindrical tank with a vertical impeller. A typical flotation circuit is shown in Fig. 11.4.

Flotation cells are built in multiple units and the flow of the pump through various units is tested for the best results. Common practice is to feed the pulp to several cells known as tougher, which produces a barren tailing and low grade concentrate. The concentrate is tested, some time after regrinding in cleaner cell and re-cleaner cells for final concentration.

The tailings from the cleaner cells are re-circulated back through the system or concentrated separately in additional cells. Regrinding of these middling is necessary in every ores.

Important auxiliary equipment in flotation plant includes reagents feeder and controls and weighing devices, slurry pumps, filter and thickeners for dewatering solids, recent storage and making up equipment and analytical devices for process control.

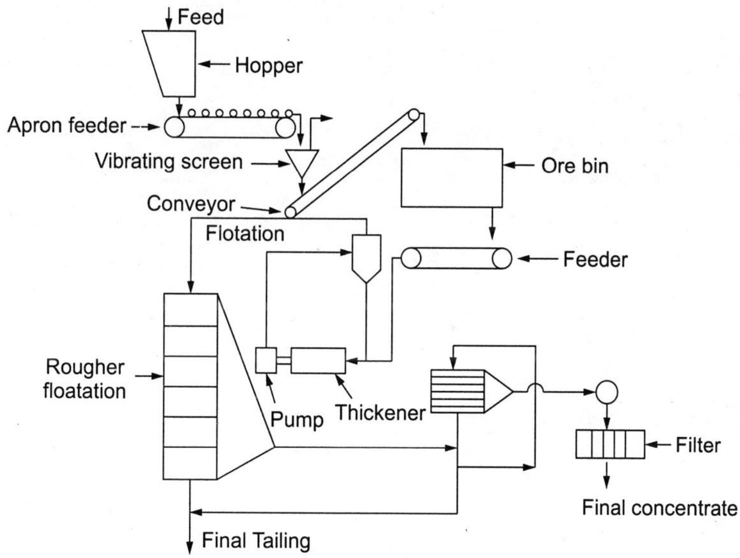

Fig. 11.4: Floatation circuit

REVIEW QUESTIONS

11.1 Define jigging.

11.2 Briefly discuss on jigging.

11.3 With a neat diagram explain the construction and working of hydraulic jig.

11.4 Define froth flotation.

11.5 With a neat diagram explain the construction and working of lab model flotation cell.

11.6 With a neat diagram explain the construction and working of D.R. Denver Flotation cell.

11.7 What are the methods of improving the flotation ability.

11.8 Briefly explain the method of improving the flotation ability by promoter and collectors.

11.9 Briefly discuss on modifier used in flotation technique.

11.10 Briefly discuss on chemical compound used as reagent to improve the flotation ability.

11.11 With a neat flowsheet explain the flotation circuit.

❑❑❑